The Thames & Severn Canal

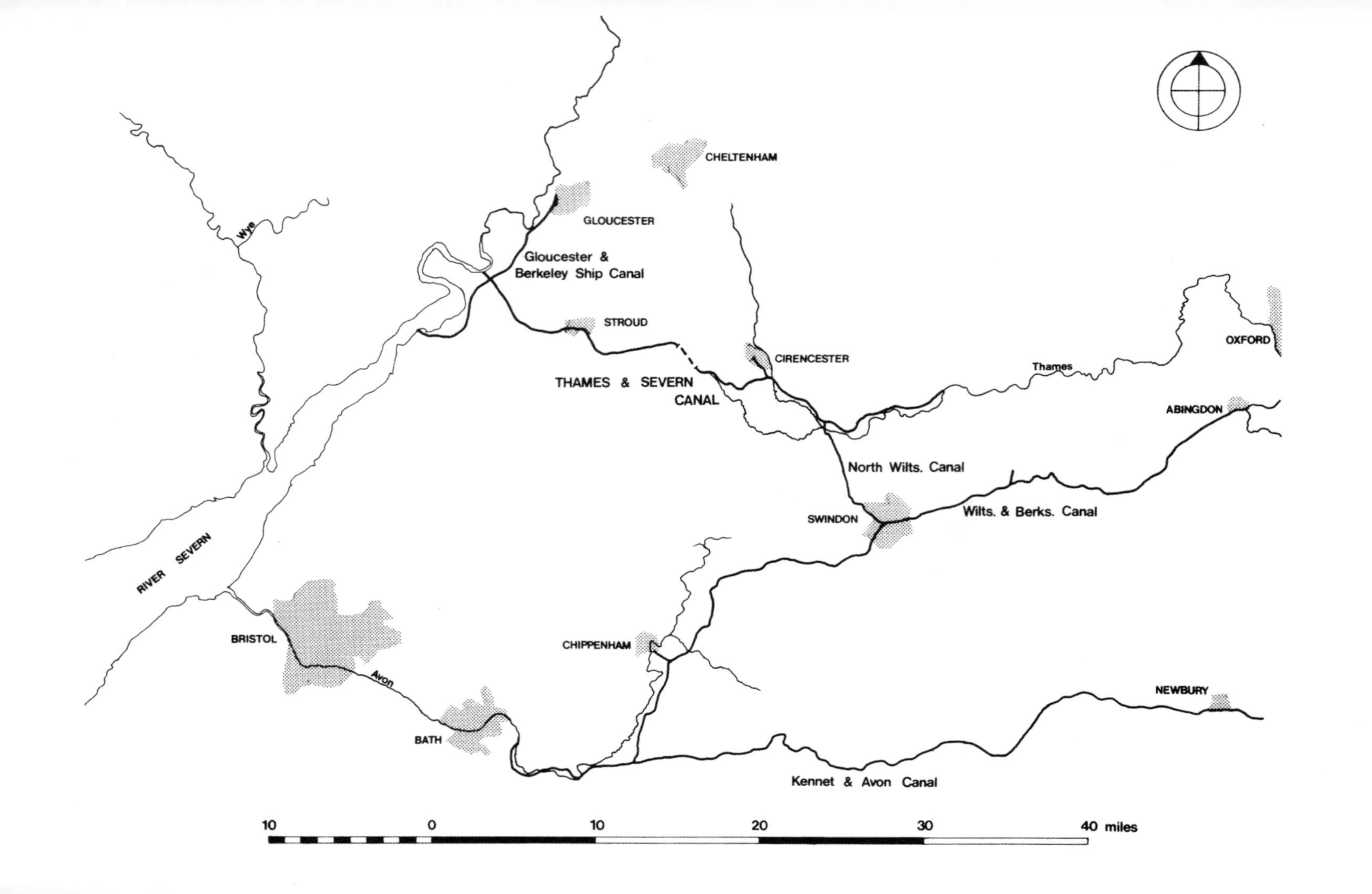
CHELTENHAM
GLOUCESTER
Wye
Gloucester &
Berkeley Ship Canal
STROUD
CIRENCESTER
OXFORD
Thames
THAMES & SEVERN
CANAL
ABINGDON
North Wilts. Canal
SWINDON
Wilts. & Berks. Canal
RIVER SEVERN
BRISTOL
CHIPPENHAM
Avon
NEWBURY
BATH
Kennet & Avon Canal
10
0
10
20
30
40 miles

The Thames & Severn Canal

Humphrey Household

AMBERLEY

Publishing history of *The Thames & Severn Canal*:

- First edition, hardback, by David & Charles, 1969
- Second enlarged edition, hardback by Alan Sutton Publishing, 1983
- Reprinted as a paperback edition by Alan Sutton Publishing jointly with Gloucestershire County Library, 1987

Other related studies

- *The Stroudwater Canal* by Michael Handford, Alan Sutton Publishing, 1979
- *Stroudwater & Thames & Severn Canals Towpath Guide* by Michael Handford and David Viner, Alan Sutton Publishing, 1984, reprinted 1988
- *The Thames & Severn Canal History & Guide* by David Viner, Tempus Publishing, 2002
- *The Stroudwater Navigation* by Joan Tucker, Tempus Publishing, 2003.

First published 1969
Second edition published 1983
This edition published 2009

Amberley Publishing
Cirencester Road, Chalford,
Stroud, Gloucestershire, GL6 8PE

www.amberley-books.com

ISBN 978-1-84868-035-7

Typesetting and origination by Amberley Publishing
Printed in Great Britain

CONTENTS

Contents

ACKNOWLEDGEMENTS

For permission to quote, I am grateful to Martin Secker & Warburg Ltd covering *Aubrey's Brief Lives* edited by Oliver Lawson Dick; to Miss F. Andrews and Eyre & Spottiswoode (Publishers) Ltd covering *The Torrington Diaries* edited by C. Bruyn Andrews; to the Institute of Geological Sciences, South Kensington, covering *The Wells and Springs of Gloucestershire* by L. Richardson; and to the London School of Economics covering *My Apprenticeship* by Beatrice Webb.

Thirteen of the illustrations are from 18th and 19th century books: coloured jacket from *Picturesque Tour of the River Thames* by William Westall and Samuel Owen, 1828; the Stroudwater committee's "Pleasure Barge" from *A Collection of Proof Prints engraved for Bigland's History of Gloucestershire*, print dated 1786; West Country barges on the Thames, and Sketches of Merchandise, from *Microcosm* by W. H. Pyne, 1806; Severn trows above Tewkesbury from *Collections for the History of Worcestershire* by T. R. Nash, 1799; navvies' tools from Rees' *Cyclopaedia*, Canal (Plates), 1819; the horse gin from *Reports on Canals etc.* by W. Strickland, 1826; Dudgrove Double Lock from *The Thames* by W. B. Cooke, 1814; the eastern portal of Sapperton Tunnel, and the windmill pump at Thames Head, from *Picturesque Views on the River Thames* by S. Ireland, 1792; Brimscombe Port from *Delineations of Gloucestershire* by J. & H. S. Storer and J. N. Brewer, 1824; Thames Head Well from *An History of the River Thames* by W. Coombe, 1794; the barge approaching Sapperton Tunnel from *A Collection of Gloucestershire Antiquities* by Samuel Lysons, 1803; narrow boats at Wallingford from *A Series of Picturesque Views of the River Thames* by W. & R. Havell, 1812.

The photograph of the bust of John Disney is reproduced by permission of the Syndics of the Fitzwilliam Museum, Cambridge; that of the portrait of Richard Potter was supplied by British Railways Western Region; those of Sir Edward Littleton and John Stevenson Salt were taken specially for me by kind permission of, respectively, Major the Hon. W. H. Littleton and the late Sir Thomas Salt, Bart.

The view of the western portal of Sapperton Tunnel before the channel had been filled is reproduced from a watercolour by permission of the Curators of the Bodleian Library. The scene of barges approaching Brimscombe Port is from an oil painting in the Stroud Museum. The plans of the wharves at Cirencester and Lechlade are based on the Ordnance Survey Map with the sanction of the Controller of H.M. Stationery Office. The drawings of the Stroud barge *Perseverance* are reproduced by permission of the Science Museum, South Kensington. The map showing the Commercial Distribution of Coal in the canal area is drawn from pp. 1830 Reports Vol. 8. The photograph of Richard Miller's banknote is from an original in the National

Westminster Bank, Stroud.

From GCRO came the sample of Samuel Smith's docketing, the sections of Sapperton Tunnel and the sketch of its distortion, plans of the ports at Brimscombe and Stroud, Charles Jones's design for a side-pond, the drawings of the carpenter's punt, the photograph of the dismantling of the boat weighing machine, and of the trow at Brimscombe Port early this century.

The photograph of the interior of the tunnel in 1955 was taken by C. P. Weaver; that of the Thames Head pumping station late in the 19th century is from the Taunt Collection by permission of Oxfordshire County Libraries; those of the cast of the seal and the halfpenny token were taken specially for me by Halksworth Wheeler, Folkestone. Those of Smerrill Aqueduct, Daneway Basin, South Cerney Lock, *The Trial* at Walbridge, *Empress* dredging, *Staunch* at Cirencester, puddling at Puck Mill, the excursion party emerging from the tunnel, are from old photographs borrowed from several different sources, including Charles Hadfield and the late P. G. Snape.

The remaining illustrations are from my own photographs.

Humphrey Household 1983

PUBLISHERS' NOTE TO THE ACKNOWLEDGEMENTS

For this 2009 edition some photographs have been replaced, some are from the author's original collection and the remainder are from the David Viner collection. Where possible we have included images known to be taken by Humphrey Household, during his own field research.

ABBREVIATIONS

BM	British Museum
BRL	Birmingham Reference Library
BTR	British Transport Records
DNB	Dictionary of National Biography
EHR	Economic History Review
GCL	Gloucester City Library
GCRO	Gloucestershire County Record Office
GJ	Gloucester Journal
JHC	Journal of the House of Commons
JHL	Journal of the House of Lords
NED	New English Dictionary
PP	Parliamentary Papers
PRO	Public Record Office
SWN	Records of the Stroudwater Navigation, now in the County Record Office
VCH	Victoria County History

FOREWORD TO THE FIRST EDITION

Of all the canals which might still be open if the present popularity of pleasure cruising had come thirty years earlier, I most regret the Thames & Severn. I have no special liking for voyaging through tunnels, but to have climbed slowly up the Golden Valley from Stroud past Chalford to Daneway when autumn's mingled browns were on the hillside trees, and then to have leisurely crossed the Cotswolds from Tunnel House inn, beneath the Foss Way, over Smerrill aqueduct, past Ewen to Cirencester and South Cerney where I live, and where my wife bathed in the canal locks as a girl that would have been a joy indeed.

The voyage is now impossible, but here in Humphrey Household's book is consolation. For him, the Thames & Severn Canal has been a lifetime's preoccupation and pleasure, and the result of his research has long been awaited by all those who knew the extent of his studies. He has been fortunate in his canal, for in the County Records Office at Gloucester is preserved probably the finest set of individual canal archives in the country: not only the central minute and account books, but a great mass of ephemera of the kind that are so seldom kept, and which so vividly recreate the past when they are.

His canal has, however, been fortunate in Humphrey Household's diligence, persistence, imagination and range of learning. The full fruits of his work lie in his great thesis in Bristol University library, but here in this book is the quintessence of his knowledge, the story of the first canal to link Bristol in the south, and Birmingham also, to London, and of the men who built what was, when it was opened, the longest canal tunnel in Britain and, indeed, in the world.

I have known the canal for thirty years and Mr Household for twenty. Now I have had the additional pleasure of seeing his book through the press and into the hands of its readers. It has been worth waiting for.

Charles Hadfield
1969

FOREWORD TO THE SECOND EDITION

Two of life's pleasures are to look back and to look forward. When Humphrey Household asked me to write a new Foreword to the second edition of *The Thames & Severn Canal*, with its useful additions to the text and double the number of illustrations, he enabled me to do both.

I re-read the old book, and concluded that it must surely be as good a history of a single British canal as has been written. Then I reflected that since 1969, when it was published, both the old canal and the book have been fortunate. The canal has gained the active Stroudwater-Thames & Severn Canal Trust, dedicated to its preservation and restoration; and in 1983, two hundred years after the authorising Act of Parliament, and long after the first edition has been unobtainable, lovers of the canal will be able to read and enjoy the second edition of Mr Household's book.

Charles Hadfield
1983

FOREWORD TO THE THIRD EDITION

Two decades is far too long out of print for a book first published forty years ago and regarded by one of the country's foremost canal historians to be "as good a history of a single British canal as has been written". Charles Hadfield was right in this assessment in my view and Humphrey Household's study of The Thames & Severn Canal has long been widely regarded as a standard work of reference, not least for the scholarship which lay behind it and in recognition of the many years the author devoted to his subject.

Other books may have appeared subsequently on the history of the Cotswold canals, but each author has acknowledged how much was owed to their 'Household', meaning both the man and his book. Indeed, the word itself seems to have become synonymous with the Thames & Severn canal.

One of my treasured possessions is a copy of the 1983 hardback edition, joyfully (and speedily) snapped up in my local Oxfam books and music shop a few years ago. It is Humphrey Household's gift copy to Charles Hadfield, who was of course one half of his original publishers as David & Charles. In it is written "To Charles Hadfield, in gratitude for your help and encouragement over a period of more than 35 years", and signed off "November 1983".

This statement and length of time take us back immediately to another lost world, to those post-war years when canal dereliction was the norm and restoration, as the leisure amenity we now see and enjoy, was something for the future. There was not even a glimmer of hope for the thirty-seven or so miles of canal which crossed the Cotswold watershed between Severn and Thames, via the valley of the river Frome above Stroud. Interestingly the first real campaigning for restoration only got under way in the very late 1960s and early 1970s, at about the same time as Household's book first appeared.

So, did he favour restoration and did his book act as a catalyst and inspiration? I asked him the first of these questions at a public lecture in Stroud a few years later and he was gracious in reply, supportive of the interest and energy by then much more clearly focussed around the campaign, but also very mindful of the long history of earlier restoration which had fallen upon various public bodies in Gloucestershire and not least upon Gloucestershire County Council since the late 19th century. As a true historian, well versed in his sources, Household could not divorce himself from the evidence of the burden of past expenditure so obvious in the records.

As to a catalyst and inspiration, his book was certainly that and I well remember buying my copy on the day it was published in 1969, for the princely sum of 50*s*. It tells an inspiring story, of course, one which has sustained public interest throughout the history of the Thames & Severn and no less today. This new edition will attract a new generation to that cause.

Humphrey Household welcomed the opportunity of a second edition in 1983 to expand his text, add new photographs and contribute a new introduction which sets the scene of his lifetime's work. There he explained his long pre-occupation with the subject, as part of his wanderings in his native Gloucestershire, latterly accompanying his father who was one-time Director of Education for the county. These wanderings by car and on foot included "walking every inch" of the canal line and battling with the more densely overgrown sections of towpath.

His career took Humphrey away from Gloucestershire "more than forty years ago" (as he recalled in 1983), which makes the body of research he was able to undertake all the more remarkable. Perhaps, as is often said, absence does make the heart grow fonder. His earliest Thames & Severn record photos date from the 1920s, at a time when there was life still in its western reaches in the valleys, linked west of Stroud to the Severn via the Stroudwater Canal.

Researching in detail the history of any one particular canal relies heavily on surviving archival evidence, and in this regard Household knew what he was facing. He learnt that the Thames & Severn's archives were housed in the Gloucestershire Record Office (and remain there to this day, now Gloucestershire Archives), and from being initially overwhelmed by the quantity and extent, he moved into a 'deep study' which lasted many years. He had the benefit, as Hadfield reminds us, of 'probably the finest set of individual canal archives in the country'.

This process of research which led eventually to his 1969 book took many years of diligence, principally focused upon his 1957-58 thesis entitled *The Thames & Severn Canal: life and death of a canal,* awarded a Masters of Arts degree by the University of Bristol. He had certainly earned this accolade and the thesis, now in the University's Library, remains the backbone for any subsequent research in this field. He lectured to the Newcomen Society in January 1950 on the early engineering of the Thames & Severn and later published a paper on the canal in the *Journal of Transport History* in May 1966, as his research came towards fruition.

The thesis of "inordinate length" needed to be reduced in order to create a manageable book and this could not have been an easy task. The result, when published, joined what is now seen in retrospect as a distinguished family of transport history titles produced during a period of memorable publishing activity by David & Charles, winning the accolade already noted from Charles Hadfield, its canal doyen. Humphrey's 'diligence, persistence, imagination and range of learning' had, he said, richly borne fruit.

The second edition in its turn came from another well-established Gloucestershire resource, the publisher Alan Sutton, whose eponymous company produced its own distinguished family of transport history titles, including some others from Humphrey

Household; see *Gloucestershire Railways in the Twenties* (1984) for example. His second edition gave its author a chance to update and expand his story, in the bicentenary year of the passing of the Thames & Severn Canal Act in 1783.

Humphrey was able to add more historical information and to bring the growing restoration story up to date, recalling the feasibility studies and assessments which laid the groundwork of much that was to follow, and delighting in the re-watering of the King's Reach between the restored Sapperton tunnel portal at Coates and the delightful (and listed) stone bridge carrying the country road to Tarlton over the canal. This remains one of those sections much visited, and acts as a constant reminder of the long historical saga which created and maintained this fascinating route across the Cotswolds.

David Viner
Cirencester
November 2008

INTRODUCTION

I was born in Cheltenham and brought up with, as it were, 'one foot on the Cotswolds', for from an early age I went on long hill walks with my father. Later I covered more of Gloucestershire on a bicycle, and still more by car, first as a very small passenger, then as companion during school holidays, and finally as driver, when my father, who was Director of Education for Gloucestershire, visited schools to the furthest confines of the county. Although it is now more than forty years since I lived in it, that early familiarity with the diverse characteristics of hill, vale and forest has ensured that Gloucestershire remains beloved by me above all other parts of Britain. So, I knew well the country traversed by the Thames & Severn Canal long before I began to study its archives and write its history; indeed, my earliest photograph of it was taken sixty years ago. Delving into the story of the canal began about fifty years ago for personal satisfaction, because among my many interests, the history of transport in various forms has been the most enduring. And so it was that when I learnt later that the Thames & Severn archives were in the Gloucestershire Record Office, I went out of curiosity to see what was amongst them. Their extent almost overwhelmed me, for there are not only the minute books, account books, deeds, maps and plans, but also a collection of notes, odd scraps of paper, and informative letters which illuminate the dry official statements with a wealth of detail. These revealed so rich a store of material bearing on the construction, operation and management of a canal that that cursory look became deep study lasting many years.

The background had to be filled in: first, by walking every inch of the line above ground, camera in hand, no easy task in those parts where the towing-path had become densely overgrown; second, because like James Perry in the closing years of the 18th century, I was "handling Subjects of the most dangerous Kind", there were frequent talks with Charles Hadfield whose knowledge of canals is so extensive, and many hours spent at the Institution of Civil Engineers who so kindly made me free of their valuable and beautiful library where I learnt to interpret the engineering data. There were repeated visits to the British Museum whose library contains the fragmentary evidence of that extraordinary 17th century preoccupation with the Thames-Severn link as well as much solid evidence of early engineering practice. To all these, my debt is great.

There were others who, having known the canal intimately in its working days, gave me a great deal of patient help and explanation: I pay my respects to the memory of C. W. Hawkes, sometime principal of Brimscombe Polytechnic housed in the old canal company's headquarters, to P. G. Snape, clerk to the Stroudwater Navigation, to Billy Stadden and Gleed of Chalford, both of whom had worked on the canal in their earlier days.

Detailed examination of the records would never have been carried out but for the encouragement of the County Archivists, Roland Austin and his successors Irvine Gray and Brian Smith, and the help they and their staff gave me, kindly accepting my presence over many weeks.

Much material was drawn from other sources than those already mentioned and for this I owe thanks to the Librarian and staff of the Gloucester City Library, to the authorities of the Bodleian Library, to the Curator of the British Transport Historical Records, and amongst others to the public libraries of Birmingham, Wolverhampton, Stafford, and Folkestone (whose staff borrowed obscure volumes from distant places on my behalf).

Of course there remain gaps I would have dearly like to fill, particularly with illustrations of the vessels built at The Bourn Yard for the lower Severn trade and of personalities such as the manager John Denyer, but diligent search failed to reveal any.

The story of the canal is one of exceptional interest. Talked of at least as early as the first decade of the 17th century, it was the first trunk waterway ever to be proposed in this country. A Bill for it was before Parliament in the reign of Charles II. Thereafter it was so frequently discussed as a national, not a local, project, that when a company was at last formed to build it, the promoters were unusually fortunate in meeting with widespread support and scarcely any opposition. In building the Sapperton Tunnel, the proprietors carried out the most ambitious engineering feat of the day. Owning and operating their own boats far beyond the limits of the canal, the company ranked for a time among the largest corporate employers in the country. Suffering under financial strain, they made one of the earliest issues of what later became known as preference shares. Principally through the work of three remarkable men appointed while still in their twenties, the canal reached its peak of efficiency in time to face that deadly threat, the railway. When earnings nevertheless shrank, the proprietors proposed converting their canal into a railway. Twice they tried, and although their second attempt was led by a resourceful railway prince, twice they failed. There followed a period of indirect railway control, ended by transfer to a Trust which failed dismally to justify the boastful claims of its promoters. Finally, for some years the canal was managed by a county council whose amateur administrators squandered the ratepayers' money on technical problems they little understood.

The story fascinated me as it unfolded, and it gives me great satisfaction that my interpretation of it should be thought worthy of a second edition in the bicentenary year of the passing of the Thames & Severn Canal Act.

Humphrey Household
Folkestone
November 1982

CHAPTER ONE

THE BACKGROUND

The very early origin of the proposal to join the River Thames and Severn forces one to ask whether contemporaries had the means which would have enabled them to make such a navigable cut, and what were the economic factors which impelled them to attempt it.

The construction of artificial waterways is a skill of great antiquity. Irrigation channels were an essential feature of all early civilisations. The Romans brought it to Britain, cutting, for example, drainage channels which formed an inland navigation through the Fens, and an open aqueduct supplying Dorchester, Dorset, with water.[1] Roman surveying methods were highly developed, amongst them instruments for determining horizontal angles and relative heights, the tape for measuring distances, and the water-level used in conjunction with plumb-line and staff to set out level or accurately graded channels.[2]

The tradition the Romans established was never wholly lost. Early in the Saxon period leaders of the church in England knew how to draw an artificial stream from a distant spring to a monastery.[3] Medieval architects and craftsmen used level and plumb-line, founded bridge abutments on piles, reclaimed parts of Romney Marsh from the sea, and formed artificial lakes to defend castles. Millwrights and carpenters of the fourteenth century and later developed the versatility of engineers and established a great tradition, for when they contracted to build mills driven by the overshot waterwheel, they had to work in timber, masonry, earth and clay in erecting the millhouse, fashioning the machinery, building the weir, making a watertight mill-leat and millpond, building mill-race and tumbling-bay able to withstand the pounding of escaping water.[4]

Rivers were also used extensively for transport, so the interests of boatmen frequently conflicted with those of millers who maintained weirs, even though navigation often depended on impounding water here and there. On navigable rivers, part of each weir was removable, forming a 'flash-lock', but riding a boat down the flash of water so released was dangerous, ascending it was a trial of strength, and the waste of water was prodigious. Obviously it could play no part in forming a navigation across a watershed. Effective junction of two rivers had to await introduction of the pound lock, consisting of two pairs of gates enclosing a chamber wherein a boat floated while the water level was adjusted to equal that of the reach above or below. This brilliant invention, wherever and whenever conceived, was greatly developed and improved by Italian engineers of the fifteenth century, and in the sixteenth its use spread to other European countries, England amongst them: three were built on the Exeter Canal about 1564.[5]

By the end of the sixteenth century, therefore, the skill to construct an artificial waterway was available, and as the joint-stock company had evolved, the means of financing one was also available.

So much for the means. What, then, was the need?

Late in the sixteenth century and throughout the seventeenth, there was tremendous progress in science and technology in England. Yet transport depended on roads often unpassable in winter, on small boats using rivers only partly or seasonally navigable, and on ships and barges plying in coastal waters and on deep rivers. These last were the cheapest and most effective means of carriage, and industrial expansion was therefore greatest in the areas they served. But as many parts of the country suffered from a shortage of fuel because woodlands were being used up faster that they could be replanted, there was a great expansion also of mining in coalfields close to sea or river, and waterborne coal was burnt instead of wood in domestic fires and an increasing number of industrial processes.

Nowhere was the need of fuel more pressing than in London, whose growth in the seventeenth century was phenomenal: doubling in size, it became some twenty times as large as any other city in England and housed almost a tenth of the population of the country. For supplies of fuel, food and manufactures, London depended mainly on the east coast trade. But the coasters had their hazards: fog and storm, reefs and rocky headlands unmarked by coastal lights, pirates plundering even in home waters, and the threat of enemy fleets ,which in time of war suspended voyages altogether.

The West and Midlands were rich storehouses of food and minerals. Their mining and manufacturing were developing fast. Some of their products already reached London by the dangerous voyage round Land's End, by land across the short watershed between the Severn and Thames, and in the western wagons which crept up the downland slopes on their way from Bristol drawn by a score of horses or oxen. It was obvious that a canal between the rivers would be more secure than the sea passage and cheaper than land carriage. Moreover, such a canal would pass through one of the most thickly populated areas in the kingdom and would benefit not only London but Bristol, then the second port and, by the end of the century, the second city in the country.

As the seventeenth century wore on, the Severnside collieries in Shropshire took their place amongst the most important in Britain, and coal carried fifty-six miles down river to Tewkesbury sold at about a third of London prices.[6] The greatest attraction to the advocates of the Thames-Severn junction was therefore the prospect of supplying London with coal and breaking the hold of the Tyneside coal-owners on the metropolitan market.

REFERENCES

1. Coates, *The Water Supply of Ancient Dorchester*, *Proceedings of the Dorset Natural History & Antiquarian Field Club*, 1901, vol. 22, pp. 80-90. The course of this aqueduct, 12 miles long on a very gentle gradient, winds along the side of the valley at an ever increasing height above the River Frome in a manner very similar to that of Brindley's canals.
2. Walters, R. C. S., *Greek and Roman Engineering Instruments*, *Transactions of the Newcomen Society*, 1921-22, vol. 2, pp. 45-60. Gunther, R. T., *Early Science in Oxford*, 1923, vol. 1, pp. 331-3. Kirby, R. S., and Laurson, P. G., *The Early Years of Modern Civil Engineering*, 1932, p. 20.
3. St. Eanswythe, First Abbess in Folkestone AD 630, traced the contour course some 2 miles long of the Town Dyke which survived until recent times (see Woodward, *The Parish Church of Folkestone*, 1894, p. 8). Salzman, L. F., *Building in England*, 1952, pp. 268-70, gives examples from the 10th, 12th and 13th centuries.
4. See Carus-Wilson, E. M., *An Industrial Revolution of the Thirteenth Century*, ECH 1941, vol. 11, pp. 39-60. Salzman, op cit, Appendix B, gives examples of such contracts.
5. Willan, T. S., *River Navigation in England 1600-1750*, 1936, p. 89. Vernon-Harcourt, L. F., *Rivers & Canals*, 1896, vol. 2, p. 203.
6. Nef, J. U., *The Rise of the British Coal Industry*, 1932, vol. 1, pp. 65, 96.

CHAPTER TWO

THE FAVOURITE PROJECT

Direct evidence of the earliest proposal to cut a navigation between Severn and Thames is lacking, but early in the eighteenth century a reliable source stated that "Mr Hill and Mr Rowland Vaughan were said to design this in Q Elizabeth's Time".[1] It is likely enough, for in 1610 Thomas Procter published a pamphlet urging improvement of highways and rivers and widespread construction of artificial navigations, and in this he mentioned the linking of Severn and Thames in a way that assumed his readers' familiarity with the idea.[2]

The problem facing any projector was how to cross the formidable escarpment which dominates the vale of Severn on the east and separates it from the Cotswold plateau whereon the Thames and its tributaries rise. There are three prominent gaps in this escarpment, each of which had its own peculiar advantages. The most southerly, the Avon valley, offered the most direct line between London and Bristol and the easiest passage across the watershed; the most northerly, the Chelt valley, the shortest route to the industrial areas of the middle and upper Severn; the middle one, the Golden Valley of the Stroudwater or Frome, a possible compromise as well as the trade deriving from its own industries. Two other routes lay further north where the Evenlode and Cherwell valleys approach the Vale of Evesham from the south, but both led to the Warwickshire Avon, a roundabout route from Severn to Thames. In course of time, all but the Evenlode valley attracted attention. Through study of the map, the distinguished astronomer and mathematician, Henry Briggs, became interested in the Avon valley route to Bristol, and between 1619 and 1630, while professor of astronomy at Oxford, he rode over on horseback to examine the watershed. From what is recorded, it can be deduced that he surveyed the country and estimated the cost of cutting through it. "Not long after this he dyed", but not before he had discussed his plan with a younger man whose enthusiasm was kindled to some purpose.[3] A little later, a Mr Hill, who might have been Rowland Vaughan's Elizabethan associate, noted the possibilities of the Stroud route as well, and in May 1633 petitioned Charles I, offering to find the best route between Severn and Thames, and to estimate the cost of a navigation cut.[4]

Eight years after Hill, a remarkable character, John Taylor the 'water-poet', gave the proposed junction greater publicity by taking a boat up the Thames and Churn almost to Cirencester, carting it across the watershed, and, armed "with a hatchett", forcing his way down the Frome to the Severn. Though he wrote exuberant descriptions for a ready market, he had a serious interest in transport and was a confirmed believer in the value of inland navigation.[5] Many of his contemporaries felt the same. There were attempts to improve several rivers, and parliamentary commissioners were ordered to consider "how Rivers may be made more navigable".[6] The time was therefore ripe for the most persistent of the seventeenth-century advocates to put forward his proposals for improving and joining rivers, in particular his favourite project for a cut between Severn and Thames. This was Francis Mathew, a gentleman of Dorset, the young man whose enthusiasm had been kindled by Briggs, and the author of three pamphlets published between 1655 and 1670.[7]

In the first of these, dedicated to Oliver Cromwell whose support he claimed to have won, he stated that he had surveyed the country between Avon and Thames, but the antiquary and gossip, John Aubrey, who knew him, remarked that it was "an ill survey",[8] and it is indeed clear enough that at this stage Mathew relied largely on the ideas of Briggs. His plan was to unite the two rivers by a cut three miles long, which is the distance between the Avon a little above Malmesbury and the headwaters of the Swill brook which joins the Thames at Ashton Keynes, streams so inadequate that Mathew evidently failed to realise the magnitude of the problem facing him. But as the years went by, his scheme matured under the influence of others, and he finally proposed to cut from Malmesbury to Lechlade, nineteen miles as the crow flies.

The navigation was to be used by billanders, flat-bottomed craft of thirty tons, towed by horses, and as the belief that watermen were dishonest was current even then, he suggested a delightfully picturesque way of restraining them: "These Billanders are to sayl up the River, every squadron by its self, having each his Admiral, and Rear-Admiral, carrying their Flags of proper Colours, none of the said Squadron sayling before his Admiral, nor any behind his Rear-Admiral; and this to prevent disorder as they pass through the Country". The opening of the Avon to navigation would, he wrote, make it possible for "the Coal-pits in Kings-wood neer Bristol, and if they suffice not, those Colliaries upon the Severn side, to furnish that Country so much destitute of Fuel up to Malmsbury and Calne".[9] When he again put forward his proposals in 1670, he stressed the advantage of making these supplies of coal available also in London, which, during the Dutch Wars, had been brought "near to the hazard of an Insurrection" by "an incredible want of Coal".[10]

While Cromwell was still Lord Protector, Mathew was joined by Thomas Baskervile, who was reputed to be "well skilled in the art of conveying water"[11] and who included a few useful details in doggerel verse.[12] Cricklade, he wrote, was then

"the highest station,
By famous Tems for Navigation,
But when th 'tis joyn'd with Bath Avon,
Then row your wherrys farther on,
ffor Baskervile, Matthews were Projectors,
how to do it since Lords Protectors,
Who did conclude, Sixty Thousand Pound,
Would throughly open, each river ground,
ffor by power of Lockes, Rains, & ffountains,
They'l make Boats to dance, upon ye mountains.
But further yet to ease your mindes
How these great works, were then design'd,
Here read their Book, there you will see,
'Twas possible, such things might be".

That book, could it be found, would be very interesting.

Cromwell certainly showed interest and is said to have offered £20,000, one-third of the cost, on behalf of the "Navy-Office",[13] which is likely enough because the Admiralty was then looking to the Forest of Dean for supplies of timber and iron, and had recently spent more than £12,600 in developing the Forest ironworks.[14] But no further progress was then made.

After the Restoration, interest in inland navigation quickened, for Charles II and those who had shared his exile had seen the use the Dutch made of their waterways. Mathew seized his opportunity anew. Advised by expert surveyors and backed by influential supporters, he turned from pamphleteering to parliamentary promotion. Joseph Moxon, a celebrated cartographer, produce a map for him.[15] Sir Jonas Moore, a distinguished mathematician and surveyor, examined the supply of the Wiltshire streams and pronounced his opinion that a cut through the high ground near Wootton Bassett offered the best chance of success.[16] George Monk, Duke of Albemarle, was a declared supporter, and above all there was the King himself, who is said to have "espoused it more than any one els"[17] and to have granted Mathew "a Private Seal for the Cut",[18] which suggests formation of a corporate body.

One Bill was introduced in 1662, but although read a second time in the Commons, it seems to have got no further than committee. No less than three Bills appeared in 1664, showing that Mathew's enthusiasm had attracted competitors, but these likewise made little progress.[19] In April 1668, he tried again, with a comprehensive Bill which has survived in manuscript[20] "for makinge of an Inland passage for Barges and other vessells from Bristoll and elsewhere to London". Amongst the undertakers were to be Francis Mathew, "sole Inventor", Thomas Baskervile, assistant, the Duke of Albemarle, and Lord Ashley, later Earl of Shaftesbury and a member of Charles II's Cabal.

How the work was to be done is indicated only very generally in terms of the powers sought by the promoters: to cut new channels; deepen or widen the natural rivers; build "Locks Weares Sluces Turnpikes pound and damms for water"; to make towing-paths for horses and men; set up wharves, warehouses and cranes; connect the navigations with mines and quarries producing coal, iron, lead or stone by means of waterways or, roads or "footrayles of Timber for Waggons"; Commissioners, ex-officio the justices of the Peace for the respective counties, were to determine differences arising between the undertakers and landowners over the value of the property to be acquired. Compensation to proprietors of mills and weirs "for pound or fflashes of water" was to be fixed. Boats of a size previously in use upon any of the rivers were to pass freely to their accustomed limits. Rates for carriage of goods — those specifically mentioned were corn, wood, salt, timber and coal — were not to exceed half the average for land carriage, according to season, for the past three years. And, quaintly enough, "right of piscarye" in the rivers was reserved to landowners and forbidden to the undertakers, their servants and watermen. Although the Bill was drawn up in ambitious terms, covering not only the routes named by Hill thirty-five years earlier but also a number of other navigations far removed from Severn and Thames, there is no doubt that the prime intention of the undertakers was to establish communication between London and Bristol. This was the first of the various navigations mentioned in the Bill. It is the only one for which the estimate of cost has survived, and it was the only one to which later writers referred, as they did, for more than a century. Yet, despite the influence of the promoters, the Bill obtained only a first reading.

Contemporary with Mathew, and of like mind though greater skill, was Andrew Yarranton, who in a book published in 1677 showed that he was well aware of Mathew's defeat nine years earlier, and, no doubt for that reason, proposed, not a physical junction of the rivers, but to make the Cherwell navigable from Oxford to Banbury and the Oxfordshire Stour to its confluence with the Warwickshire Avon, relying on land carriage between the two over "good hilly sound dry land" which would bear wagons at any season of the year.[21]

The most far-sighted seventeenth-century exponent of inland navigation was, however, no projector but the author of a paper read to the Royal Society in 1675.[22] In this, Sir Robert Southwell broke clean away from the obsession of his contemporaries with river improvement and expressed an opinion which did not win general acceptance until more than a century later, namely that "in most cases it were better and cheaper to make new channels". Southwell's principal object was the same as Mathew's, to increase the supply of coal to the London market. To do this, and also promote the flow of general trade and the carriage of passengers, he envisaged the cutting of "one thousand miles of new navigable channels" connecting London with some inland coalfield and perhaps twenty other centres in the country. Relying, perhaps too hopefully, on the labour of "all malefactors not deserving death, and all idle persons", he supposed the work might be done at a cost of 4d. a cubic yard "where the ground is fit to hold water, and is not too contumacious by reason of rocks and other impediments".

In 1717, using the new word canal[23] instead of the hitherto customary cut or passage or channel, Dr Thomas Congreve, a Wolverhampton physician, published proposals for a Thames-Severn canal in conjunction with a Trent-Severn canal, prophetic of the association which eventually gave it birth. Although primarily concerned with the Trent-Severn junction, he outlined the history of the Mathew-Baskervile proposals, and suggested a very similar route between Thames and Avon by which "A Boat might pass from London to Bristol in ten Days".[24]

For more than a hundred years then, there had been propaganda in favour of canal building in England by men who wanted to emulate the successful navigations on the continent. Meantime, between 1666 and 1681 the French had built the greatest canal of the age, the Languedoc, opening communication between the Mediterranean and the Atlantic. Inevitably the question arises: why was no similar work executed in Britain? It was certainly not due, as has often been alleged, to a lack of technical skill.

Surveyors had a variety of instruments: the wire-line for measuring distances, or even better, for most of the period the familiar chain; the circumferentor for reading horizontal angles, the quadrant for vertical, the theodolite for both, and the ancient water-level, all of them with open sights only.[25] In 1674, however, William Leybourne described the far superior level with telescopic sights below a tube with bubble in spirit, and by 1701 this was being used by engineers and surveyors engaged on urban water supply, river improvement, and draining marshland.[26] It was the forerunner of the instrument extensively used by the canal builders of the second half of the eighteenth century, and in the form of the Dumpy Level remains part of the surveyor's equipment to this day.

Engineers had long been actively engaged in a great variety of waterworks. Of the simpler types, knowledge and practice were widespread, for it was water-power that drove the ubiquitous corn mills, the fulling mills of the woollen industry, the bellows, hammers and slitting-mills of the ironworkers, and the majority of mine pumps. Early in the seventeenth century, the New River had been built to supply London with water, and in land drainage not only the Dutchman Vermuyden, but also the Englishman John Perry, had been engaged in successful reclamation schemes.[27] Meantime, inventors were busily devising machines to help cut, dig, cleanse, dredge, and make rivers navigable,[28] and after the Restoration there had been a second and a third spate of river improvement Bills before Parliament. Much had been achieved, and travellers were pleased to note the results, Defoe, for example, commenting on many rivers recently made navigable.[29] The available evidence, indeed, suggests that there had been an increase of almost 70 per cent since 1660, and that by 1724 there were about 1,160 miles of river navigation in England.[30]

Technical skill was well developed in other directions, especially mining. Timber railways had been laid from several English collieries very early in the seventeenth century.[31] The boring rods used later by civil engineers to reveal geological structure had been evolved for mine prospectors about the same time. Mine galleries and drainage adits had been driven a mile or more underground. In sinking shafts, raising coal by horse-gins, ventilating collieries, and using gunpowder to blast ore-bearing rock in underground galleries, British mining engineers had already developed the technique used in driving navigation tunnels a century later. Furthermore, the two most intractable technical problems of the day were solved in the early years of the eighteenth century: how to use coal instead of wood in iron smelting, achieved at Coalbrookdale on Severnside by Abraham Darby I; and how to drain deep mines, achieved by Thomas Newcomen's invention of the atmospheric steam engine.[32]

The knowledge and skill of the various seventeenth and early eighteenth century projectors are less easy to demonstrate. Books and pamphlets primarily intended to show the economic advantages of inland navigation obviously say little about engineering. Nevertheless they and other sources do reveal some details. Rowland Vaughan, 1559-1631, was a pioneer of irrigation in England and knew a good deal about making artificial waterways on which he used twenty-foot boats to carry compost, earth and produce.[33] Procter displayed sound knowledge of road formation and claimed experience of river improvement. Hill was evidently a competent surveyor each of his three routes was subsequently followed more or less closely by a canal. Mathew almost certainly lacked practical experience and depended on the technical skill of others, but under their influence his ideas expanded, and his knowledge, however he acquired it, was remarkably extensive and in some respects esoteric. Only from the coalfields of Shropshire, Nottinghamshire or the north could he then have learnt of the timber wagonways he proposed to use. His criticism of a lock between Bath and Bristol, built on the weir and exposed to damage by the river when in flood; his instance of a Flemish lock near Ypres as an exemplary model — the very one described by an engineer a century later as "the finest Lock in Europe";[34] his recommendation that each lock should be placed in a cut separated from the course of the river and curved like a bow to deflect the force of the stream — all these are incontrovertible evidence of extensive technical knowledge and discernment. Further, his advocacy of horse towing was not only in advance of contemporary practice in Britain, but of much later custom on the Rivers Thames and Severn and some canals.

Baskervile, as we have seen, was reputed to have practical skill "in the art of conveying water". Yarranton had surveyed the rivers Thames, Severn, Trent,

Dee and Salisbury Avon, and claimed to have made the Warwickshire Avon navigable to Stratford, and to have improved the River Stour from Stourbridge to Kidderminster. Nor does his own testimony lack support, for Nash, the eighteenth-century historian of Worcestershire, remarked that it was not long since some of the boats made use of in Yarranton's navigation [of the Stour] were found".[35] Yarranton's enemies attacked him for alleged political and religious intrigues and for exceeding his estimates, but not for lack of technical skill.[36]

Southwell and Congreve, who had no axes to grind, gave more engineering data than the others. Congreve's detailed exposition of distances, levels and streams between Trent and Severn shows that a survey had been made and suggests that an engineer's report had been obtained, as does his interesting proposal that the summit in the moors near Wolverhampton should be supplied by building dams to create a huge reservoir fed by seven brooks. This was probably inspired by the description of the reservoir on the Languedoc Canal published in the transactions of the Royal Society. It was the solution of a problem formulated by Southwell, that of water supply where direct intake from streams was inadequate, but it was a solution very probably intended by Mathew and Baskervile in applying for power to "bild and sett . . . damns for water". In this respect, indeed, late seventeenth and early eighteenth-century theory appears to have been ahead of later engineering practice, as until the 1790s very few canals depended upon supply from reservoirs.[37]

Both Mathew and Southwell were aware of the impact of geology. Both Procter and Southwell proposed cuts of similar width, sixteen or sixteen and a half feet, Southwell definitely stating that this was adequate for only one vessel to pass at a time: a century later fifteen feet was the width adopted for parts of the Thames & Severn Canal similarly restricted. Southwell suggested that the depth of water should be five feet and this was the depth chosen for all but the summit of the later Thames & Severn Canal. Mathew proposed the use of vessels with a maximum draught of three and a half feet, and the ratio between Mathew's draught and Southwell's depth is, whether fortuitous or not, very close to that advocated by engineering science and practice in the nineteenth and twentieth centuries.[38]

As for estimates of cost, Procter reckoned £250 a mile, and Southwell's figure works out at about £300. Each appears to have had only straightforward excavation in mind, excluding locks, weirs and bridges — Southwell indeed suggested that "to comprehend all accidents" the cost might be ten times as great. This was a liberal allowance, as the estimates for the Thames & Severn Canal in 1782 show that provision of those essential features multiplied the cost of open cutting by four. Four times Southwell's figure was rather more than the expense of the first part of the Bridgewater Canal.[39] As we have seen, Mathew and Baskervile estimated £60,000 as the cost of improving the river navigations and making an artificial cut from Malmesbury to Lechlade, nineteen miles as the crow flies, perhaps twenty-five by a canal. At £1,200 a mile, such a cut would have cost £30,000, leaving a like amount for river improvement, principally of the Avon, as the Thames was then considered navigable below Lechlade.

Such evidence as there is therefore not only confirms Mathew's claim that there was "in this Island, Sufficiency of all Sorts of Provisions for undertaking" the works he had in mind,[40] but also indicates that British engineers of the period, so far from lacking the skill to build a Thames-Severn canal, were fully aware of what ought to be done, how to do it, and how to estimate the expense. The reason for their failure to turn theory into practice must be sought in the realm of finance.

It is quite certain that contemporaries would have seen the financing of a Thames-Severn canal as a formidable, even perhaps a terrifying, problem. They could not have failed to be aware of the notorious unreliability of estimates as several contemporary examples of river navigation schemes had run away with wholly unexpected sums of money.[41] Moreover, strange as it may seem nowadays, a capital of £60,000 for a joint-stock company was a very large sum in those days. When Mathew's Bill was before Parliament, there were only some half-dozen companies whose capital exceeded that amount, and even thirty years later, after a tremendous burst of promotion, there were probably not more than ten whose capital value was £60,000 or more.[42] Furthermore, these ten were engaged in foreign trade, banking, manufacture and London water supply, all of them activities which promised a more assured return than a project of inland navigation. One may fairly safely assume, therefore, that the "foolish Discourse at Coffee-Houses" which, said Yarranton, "laid asleep [Mathew's] design as being a thing impossible and impracticable",[43] was the bandying about of doubts whether it was possible to raise such a large sum of money for the purpose, whether that sum would suffice for the work, and whether the prospect of the navigation earning an adequate return on the capital was sufficiently attractive. So far as the carriage of coal was concerned, the doubts were certainly justified, for even when the canal system had been fully established, the quantity of coal carried to London by the inland navigations was very small in comparison with that brought by sea.[44]

With the examples of state enterprise in other lands in mind, Mathew himself in 1655 held the opinion that the cost of his Thames-Severn project was "too great . . . for any private man or Corporation to lay out",[45] and although he was evidently persuaded to take a different view meanwhile, by 1670 he had again reached the conclusion that it was "an Undertaking so Heroick" that only the state could carry it through.[46] In this country, however, the Crown was in no position to engage in such a venture in the seventeenth century. Nor indeed was canal navigation so essential for the expansion of trade in England as in other lands, for this sea-girt island has an exceptional proportion of coastline to land-mass, and in the seventeenth century coastal shipping was already well developed.

Nevertheless, although the favourite project of joining Thames and Severn was not brought before Parliament between 1668 and 1783, it was never long without an advocate. It seems indeed that its accomplishment became a point of honour, for when the time was ripe, it was exceptional, if not unique, in meeting with scarcely any opposition from the landowners along its course.

REFERENCES

1. Congreve, T., *A Scheme or, Proposal For Making a Navigable Communication Between the Rivers of Trent and Severn*, 1717, p. 13.
2. Procter, T., *A Profitable Worke to this whole Kingdome*, 1610.
3. Aubrey's *Brief Lives*, ed. O. L. Dick, 1950, p. 38. Aubrey's entry on Briggs is devoted exclusively to the Thames-Severn project. See also DNB Henry Briggs, 1561-1630.
4. PRO SP 16/238/3, 1-17 May 1633.
5. Taylor, *John Taylor's last Voyage, and Adventure*, 1641. See also DNB John Taylor 1580-1653.
6. Willan, op cit, pp. 26-7, 69, 89-92, 94-5. *Parliamentary: or Constitutional History of England*, vol. 19, p. 315.
7. Mathew, F., *Of the opening of rivers for navigation 1656; A Mediterranean Passage by Water Between ... Lynn and Yarmouth*, 1656; *A Mediterranean Passage by water, from London to Bristol &c, And from Lynne to Yarmouth*, 1670.
8. Aubrey's *Brief Lives*, p. 39.
9. Mathew, *Of the opening of rivers*, pp. 7, 9.
10. Mathew, *London to Bristol*, p. 8.
11. Thacker, F. S., *The Thames Highway, A History of the Locks and Weirs* 1920, p. 25, quoting from a note of 1683 among the Rawlinson Mss in the Bodleian Library.
12. Baskervile, T., BM Harleian Ms 4716.f.4.
13. Congreve, op cit, p. 14.
14. Nicholls, H. G., *Iron Making in the Olden Times*, 1866, p. 42.
15. Yarranton, A., *England's Improvement by Sea and Land*, 1677, p. 64. DNB Joseph Moxon 1627-1700.
16. Aubrey's *Brief Lives*, p. 39. DNB Jonas Moore 1617-1679.
17. Aubrey loc cit.
18. Congreve, op cit, p. 14.
19. JHC vol. 8, pp. 362, 370, 546, 571, 576.
20. House of Lords Record Office, Ms Bill "prima vice lecta 14° April 1668". JHL vol. 12, pp. 221-2.

21. Yarranton, op cit, p. 65. DNB Andrew Yarranton 1616-1684?
22. Birch, T., *History of the Royal Society*, 1757, vol. 3, pp. 207-210, part of a "discourse concerning water".
23. NED quotes 1673 as the earliest literary use of Canal in the sense of an artificial watercourse to serve the purpose of inland navigation.
24. Congreve, op cit, p. 13.
25. Gunther, op cit, vol. 1, and Wolf, A., *A History of Science, Technology, and Philosophy in the sixteenth and seventeenth centuries*, 1935.
26. Leybourn, W., *The Compleat Surveyor*, third ed. 1674, p. 313. Tuttell & Moxon, *The Description and Explanation of Mathematical Instruments*, 1701, pp. 12-13.
27. DNB John Perry 1670-1732. Smiles, S., *Lives of the Engineers*, Vol. 1, 1862, pp. 69-82. Perry, J., *An Account of the Stopping of Daggenham Breach*, 1721.
28. Abridgements of Specifications relating to Harbours, Docks, Canals &c 1617-1866.
29. Defoe, D., *A Tour through England and Wales.*
30. Willan, op cit, p. 133.
31. Simmonds, J., *The Railways of Britain*, 1961, p. 1. Dendy Marshall, C. F., *A History of British Railways*, 1938, p. 5. Jenkins, Rhys, *Collected Papers*, 1936, No. 6, *Railways in the sixteenth century*. Lewis, M. J. T., *Early wooden railways*, 1970, pp. 90-104.
32. See Ashton, T. S., *Iron and Steel*, pp. 28-36, 249-252. Galloway, R. L., *History of Coal Mining*, 1882, pp. 55-6, 58, 60-1, 70, 84. Galloway, R. L., *Annals of Coal Mining* First Series 1898, pp. 159, 168, 178, 194, 244. Nef, op cit. Vol. 1, pp. 243, 353, 356.
33. Vaughan, R., *Most Approved, And Long experienced Water-Workes*, 1610. Bradford, C. A., *Rowland Vaughan, An Unknown Elizabethan*, 1937.
34. Vallancey, C. , *A Treatise on Inland Navigation*, 1763, p. 164. He stated that the lock had been built in 1643.
35. Nash, T. R., *Collections for the History of Worcestershire*, second ed. 1799, Vol. 2, p. 45, and see also Vol. 1, p. 306.
36. See "A Coffee-House Dialogue" and other tracts, BM T.3^{x} 17 and 18, and BM 1852.b.2.41 and 51.
37. See Observations on the use of reservoirs for flood waters by William Jessop in Pitt, W., *General View of the Agriculture of the County of Stafford*, 1794, pp. 39-44.
38. See Stevenson, D., *The Principles and Practice of Canal and River Engineering*, third ed. 1886, p. 14, and Minikin, R. C. R., *Practical River and Canal Engineering*, 1920, p. 112.
39. Anonymous, *The History of Inland Navigations, Particularly those of the*

Duke of Bridgwater, in Lancashire and Cheshire, 1766, p. 41.

40. Mathew, *London to Bristol*, p. 3.
41. See for example, Nash, op cit, Vol. 1, p. 446, for the story of William Sandys and the Warwickshire Avon 1635-8, Willan, op cit, pp. 69-70, for the involved finances of the River Wey improvement 1651-3, and "A Continuation of the Coffee-House Dialogue" (BM Tract 1852.b. 2.41) for a contemporary attack on Yarranton for exceeding his estimate.
42. Scott, W. R., *The Constitution and Finance of English, Scottish and Irish Joint-Stock Companies to 1720*, 1911, vol. 1, pp. 325-8, 334-6, and vol. 3, pp. 459 et seq.
43. Yarranton, op cit, pp. 64-5.
44. PP 1871 Reports vol. 18, pp. 863, 1147.
45. Mathew, *Of the opening of rivers*, 1656, p. 14.
46. Mathew, *London to Bristol*, 1670, Introduction.

CHAPTER THREE

STROUDWATER

Throughout the seventeenth century, it was the Avon valley route between Severn and Thames that attracted almost all the attention, but in the eighteenth it was the Stroudwater-Thames Head line. Although this route across the watershed was clearly far more difficult, there were sound economic reasons in its favour. Between 1690 and 1760, the woollen industry of the Stroud valleys reached the height of its prosperity,[1] but the clothiers found their trade hampered by poor roads and the heavy cost of carriage overland before cloth could be shipped from Gloucester or Bristol.[2] In addition there was the problem of supplying the thickly populated area with fuel and food. A navigation up the valley to Stroud would obviously make transport easier, but although many clothiers were ready to support one, there were others, principally those with mills below Stroud, who hotly opposed any interference with the natural flow of the streams which not only drove the mill-wheels, but also supplied water whose softness and purity had done much to make Stroud cloth famous. This deep division of opinion made a Stroudwater navigation a matter of bitter controversy for three-quarters of a century.

Some far-sighted clothiers may have thought of extending the navigation to the Thames, but it was Allen, first Baron and later first Earl Bathurst, who then gave this idea prominence. One of the pioneers of eighteenth-century landscape gardening, Allen Bathurst, enthusiastically enlarging and embellishing his estate, suggested that the junction of the two rivers under Sapperton Hill should be the crowning feature of his splendid Cirencester Park, and in 1722, while staying with his patron at Cirencester, the poet Alexander Pope wrote to a friend telling of Lord Bathurst's dream that Thames and Severn should be led to "celebrate their Marriage in the midst of an immense Amphitheatre" which would be "the Admiration of Posterity a hundred Years hence".[3]

It was a poet's romantic description, but there was more to it than that. Lord Bathurst supported a Stroudwater navigation Bill eight years later, and he alone could have been "the noble Lord of particular local influence" who, it is said, then mentioned extension of the Thames navigation to Cirencester and a link between it and the Stroudwater as a possible future development if the Bill should pass.[4] It is even possible that he had taken the advice of an engineer of unusual vision, John Hore of Newbury who prepared the Stroudwater plans and had previously been engaged to improve the navigation of the Kennet, the Bristol Avon and the Chelmer. Hore's experience had taught him that to cut artificial channels was more effective than any tinkering with the rivers themselves. For the Kennet navigation, his cuts had eliminated a great part of the river; for the Avon between Bath and Bristol, he "adopted and executed a navigation, of which . . . there then existed no model in England"; and for the Chelmer, he prepared alternative plans, one for "making the old Stream useful" and the other for "cutting thro' the higher Grounds . . . and making an entire new Navigable Canal".[5] In fact, his practical experience proved the soundness of Sir Robert Southwell's prediction made more than fifty years before.

Hore was just the man for Stroud, where any attempt to improve the river itself would have aroused implacable opposition, and moreover would have failed to accommodate the 60-ton boats which the promoters had in mind.[6] Hore therefore planned a 'New Cut' 8¼ miles long, 33 feet wide and five feet deep, with twelve locks which were to be filled by "a Pipe of Four Inches Diameter" without injury to the mills.[7] He estimated the cost as less than £20,000. The Bill, modified to appease those clothiers who opposed it, was enacted in 1730,[8] and although local opposition prevented any effective use of the powers, it is of great interest, not only because the plan was so far ahead of current engineering practice, but because the navigation actually built nearly fifty years later was based closely on Hore's plans. Furthermore, a peculiar feature of the Act had the effect of prolonging its validity indefinitely; for the commissioners nominated to supervise the exercise of the statutory powers were given authority to appoint new undertakers if necessary, and to fill any vacancies in their own ranks, and as the Act had not specified precisely how "the River Stroudwater" was to be made navigable, they were able to assume the right to interpret even this, encouraging during their long corporate existence the adoption of very different methods.

A quarter of a century later, John Dallaway of Brimscombe, who was himself one of the commissioners, put forward a much less ambitious scheme for making the Frome navigable by boats of 60 or 70 tons to Fromebridge and thence by 30 or 40 ton boats to Stroud. After gaps in the ranks of the commissioners had been filled, he persuaded them to back his plan, which had been prepared after consultation with seven surveyors, one of whom, Thomas Yeoman, was an engineer of great distinction who followed Hore not only at Stroud but also on the Chelmer. Yeoman proposed that there should be sixteen pound locks and that water should be drawn from a reservoir filled on Sundays when the mills were idle, instead of by Hore's pipe which implied a constantly running stream. Of particular interest is Dallaway's expressed belief that a canal to Stroud would bring "nearer together the Navigation of the Two principal Rivers in the Kingdom, Thames and Severn, and the opening a nearer Communication by Water between the Two chief Cities, London and Bristol".[9]

The next plan for making "the River Stroudwater navigable" was without doubt the most cumbrous system of navigation ever devised, and certainly one which could never have formed part of a communication with the Thames. Nevertheless it looked as though it would disarm the dissident mill owners, and the commissioners therefore decided to give it a trial. This was Thomas Bridge's "New Machine for Making Rivers Navigable",[10] a crane — or rather a pair of cranes — on the dam of each millpond to lift frames carrying a ton of goods between boats floating above and below the dam. Josiah Tucker, Dean of Gloucester and a leading economist of the age before Adam Smith, saw part of the system in operation and thought highly of it, considering it better than the use of pound locks, but even he admitted that others thought Bridge "a chimerical, crack-brained fellow".[11] In fact, it was not long before someone better informed demolished the dean's case entirely by pointing out that his comparison had been with out-moded locks, and that frames so limited in size would prevent the carriage of heavy articles weighing several tons — a prime function of a waterway.[12] But even Bridge's scheme did not pass unchallenged by the opposition, and fresh powers had to be sought. The amending Act was obtained without difficulty in 1759,[13] and thereafter Bridge and his partners took some eight years to complete about half the installation before they gave up "'and left their Works in Ruins"[14] — not entirely in 'ruins' however, for some part was used for years by the owners of mills alongside.[15]

Junction of Severn and Thames was a project still very much alive. Congreve's pamphlet had been reprinted in 1753.[16] The following year both Avon-Kennet and Avon-Thames routes had been put forward.[17] The outbreak of the Seven Years' war led George Perry, an engineer with the Coalbrookdale company who had made a close study of the trade and navigation of the Severn valley, to write in 1758 of the "vast advantage" an inland waterway between London and western ports would be in time of war, adding that it was a pity "union of the intermediate rivers" could not be listed "among the attempts of the like nature, now on foot".[18]

In fact, several such 'attempts' were then being discussed and shortly afterwards were certainly 'on foot'. In Lancashire, the promoters, of the Sankey River navigation had determined to build a parallel canal instead of tinkering with the stream, and the third Duke of Bridgwater had engaged the Staffordshire millwright, James Brindley, to build a canal from his Worsley Colliery to Manchester. The Bridgewater Canal was the first navigation that did not follow a river line and the first to be directly connected with a coal mine (as suggested by Sir Robert Southwell), and it was therefore a landmark in the history of British transport.

The exciting engineering triumphs of Bridley and the Duke aroused curiosity and admiration.[19] The financial success encouraged emulation. In 1766, Sir Richard Whitworth, a landed proprietor and MP for Stafford, published "The Advantages of Inland Navigation", a book which was widely circulated and had great influence. Within two years, companies were formed to build the Trent & Mersey, Staffordshire & Worcestershire and Birmingham canals, linking the midlands with the river navigations leading to the three great ports of Bristol, Liverpool and Hull as Sir Richard suggested. Brindley was the engineer to all three companies, and proposing to link this network with London — probably at the instigation of the promoters of the Staffordshire & Worcestershire Canal — he surveyed a line between Severn and Thames.[20]

Sir Richard Whitworth suggested using boats eleven or twelve feet wide, capable of navigating the tideways. However, to allay the fears of proprietors alarmed at the costs of construction, Brindley reduced the scale of his engineering on these trunk canals and built locks and tunnels taking vessels no more than seven feet wide. It was a great pity: cargoes destined for ports on the estuaries had to be transhipped to stouter craft,[21] and in modern times the limitation has proved a fatal handicap to commercial traffic on the narrow canals.

With so much enthusiastic canal promotion going on, it was not long before plans for the Stroudwater were revived. The need was greater than ever. The homes and industries of the valley had multiplied, and it was reckoned that the local consumption of coal — all of which had to be brought by wagon, or in winter on horseback — was 12,500 tons a year.[22]

Early in December 1774, a group of nine men, led by William Dallaway, John's son and himself a clothier of distinction who had served as high sheriff of the county, prevailed on the commissioners to name them as undertakers. Assuming that the powers of 1730 were still applicable to a navigation based on John Hore's plan, they set out to raise £20,000 to meet Hore's estimate of the cost. Before the end of the month they met as the Committee of Directors for the Stroudwater Navigation and appointed one of their number as clerk — Benjamin Grazebrook, a man of many parts who by providing Stroud with a water supply, had already proved himself a competent surveyor, and who in course of time was banker, partner in the Stroud Brewery, and canal and river carrier. They called in "an eminent and experienced engineer", none other than Thomas Yeoman, by then Fellow of the Royal Society and first President of the Smeatonian Society of Civil Engineers, who had been in practice at least forty years, and who, having collaborated with John Dallaway twenty years earlier, was familiar with the neighbourhood and its problems.[23] Yeoman's navigation, which was indisputably a canal, closely resembled Hore's. There were twelve locks (a thirteenth, Junction Lock was inserted in 1826), the total rise was 107 feet 2 inches above low water in the Severn, the length was 7 miles 74 chains. Dimensions in the open cutting were increased to 42 feet wide and 6 feet deep to accommodate the larger vessels by then in regular use on the Severn, 15 feet 2 inches wide and 68 feet long. He proposed to supply it from a reservoir at Walbridge, Stroud, the terminal basin.[24]

Having appointed a Mr Priddey as "Chief Engineer and Surveyor"[25] really, resident engineer — the committee met at Framilode on Severnside at the end of May 1775 for William Dallaway to lay the first stone of the entrance lock. After nearly half a century of frustration, this was clearly an occasion of celebration: there were "chearful shouts" from "a great concourse of . . . spectators" and dinner at the little inn at Framilode Passage, when toasts were drunk and satisfaction expressed. But although some were so intoxicated by the heady enthusiasm — if nothing more — that they declared the navigation was already on the way to "being speedily completed", there were others who thought "the malignant spirit of opposition" had still to be reckoned with.[26] They were right.

In fact, some of the owners of lands and mills below Stroud had already raised a fund and let it be known they would challenge the new undertakers' assumption of authority when it became clear exactly what they intended to do. After men had dug about half-a-mile of a channel parallel to the River Frome, there was no longer any doubt that the intention was no mere improvement of the river. The land cut was held by a sub-tenant, and although he, like the landowner, had given consent, the primary tenant had not even been approached. Here was the awaited opportunity, for he was an implacable opponent, and it was he who issued the challenge. Anxious to complete the entrance lock so as to prevent damage by the tides, the committee suggested testing the validity of the Act "upon a feigned issue" at the approaching Gloucester Assizes.[27] The opposition would have none of this, however, until after successfully applying for an injunction to stop the works, they were told by the London court to do just that. When the case came up at Gloucester in August, the proprietors lost on the grounds "that a Canal Navigation, as this was deemed, could not be made" under the existing powers.[28]

There was nothing to be done but apply for another Act. Yeoman and Priddey were asked to prepare fresh estimates. The new Bill was presented in November. Its passage was again opposed by the mill owners below Stroud, and also by the City of Gloucester which feared diversion of trade from the town. But the Merchant Venturers of Bristol supported it, and it was one of that city's MPs, Edmund Burke, who introduced the Bill and whose persuasive eloquence helped to get it through. Royal Assent was given in March 1776, and at last the way was clear.[29]

Work recommenced, but the Stroudwater Navigation's long history of litigation was not yet over. At Ryeford, east of Stonehouse, there is a double lock, rising 16 feet 6 inches in two equal stages divided by a common pair of gates. When the lock-pit was excavated in the side of the valley, which there rises some 20 feet above the Frome, the bank repeatedly slid down. No mason would undertake to build the walls. At last Anthony Keck of King's Stanley, one of the proprietors and "a person of great abilities" (he was architect of St Martin's Church and the Infirmary in Worcester), produced a model showing how the bank could be upheld by plates supported by timber frames, braced and tied. His plan was approved and he built the lock under contract. But in less than a year, the walls began to bulge in so menacingly that the defective brickwork had to be taken down. The committee sued Keck on the ground that one of the principal timbers was less than the size specified in the model. Seizing upon this point, learned counsel addressed the jury with great effect, repeatedly asking "Gentlemen, can a packthread string be as strong as a cable rope?", and Keck lost. "Out of 30 law-suits relative to this canal", wrote a contemporary, "no one excited such general interest", and indeed small wonder, for although the case does not seem to have been cited as a precedent in the law of contract, its influence must have been considerable. Obliged to reconstruct the lock at his own cost, Keck solved the problem by draining the unstable ground, turning in the lock wall three brick cylinders 4 feet in diameter; these served their purpose admirably and can still be seen today.[30]

It seems curious that no engineer appears to have played any part in

Stroudwater Navigation, Ryeford Lock in 1950, showing the three brick cylinders turned in the wall of the upper chamber to drain the unstable bank.

The Shallop built for the Stroudwater committee by James Martin Hilhouse of Bristol as their 'Pleasure Barge' passing Stonehouse Church. From a print dated 1786.

tackling Ryeford lock. But in fact the resident engineer's lot at Stroud was not a happy one. The committee were apt to complain of lack of progress and to issue orders directing this or that to be done. Three engineers were dismissed. Two others invited to take their place found reasons for declining. Finally the versatile and capable Benjamin Grazebrook was appointed to superintend completion of the works for a fee of £200.[31]

Lengths of the canal were put to use as completed from 17 December 1776 onwards, and by 21 July 1779 the line was ready for opening to Walbridge. The day so long awaited was clearly one for uninhibited celebration. The committee arranged a 'public breakfast' and a procession through the town and on the water, attended by the MPs for the county, citizens of Stroud, the navvies, a band, and of course the triumphant committee themselves; and for the evening they arranged a ball at the town hall. But although the public breakfast may have been free, attendance at the ball was not, and the committee even charged spectators for standing room on the towpath.[32]

For the occasion, Ralph Bigland of Frocester, Garter Principal King of Arms and historian of Gloucestershire, had made a "Genteel Present of Flaggs to the Company".[33] His son Richard was a member of the Stroudwater committee, and it must surely have been the scene on the opening day that prompted the design of an engraving prepared some years later for Richard and intended, though not in fact used, to illustrate his father's history of the county.[34] It shows a shallop dressed with flags and carrying three musicians overtaking a barge by Stonehouse Church to the evident interest. of passers-by — but one of the flags bears the date 1781, a year of no particular significance in the company's history. The shallop, however, is undoubtedly the committee's "Pleasure Barge" for which James Martin Hilhouse, the well-known Bristol shipwright, was paid £94 12s. in August 1780,[35] and which had taken part in the opening procession.

The Biglands were also responsible for the company acquiring an unusually fine seal, designed by one of the Pingo family, distinguished medallists and engravers at the Royal Mint.[36] When it was ready, Pingo sent in an account for £20 for the matrix and £7 10s. for an iron chest in which to keep it. The fee was reasonable enough (the Ellesmere Canal Company later on had to pay £31 10s. for their seal),[37] but the Stroudwater Committee told the Biglands roundly that they had no wish for the chest and could "not consent to give so enormous a Sum for the Seal". In the end the account was settled for £20 16s.[38]

There was a reason for the committee's stinginess. Inflated by the heavy legal expenses,[39] the cost of the canal greatly exceeded expectations. The Act of 1776 authorised a share Capital of £20,000 with power to raise another £10,000 by either issuing more shares or borrowing. The extra money was needed as early as the autumn of 1777, but rather than increase the share capital or mortgage the property, the proprietors undertook to find the £10,000

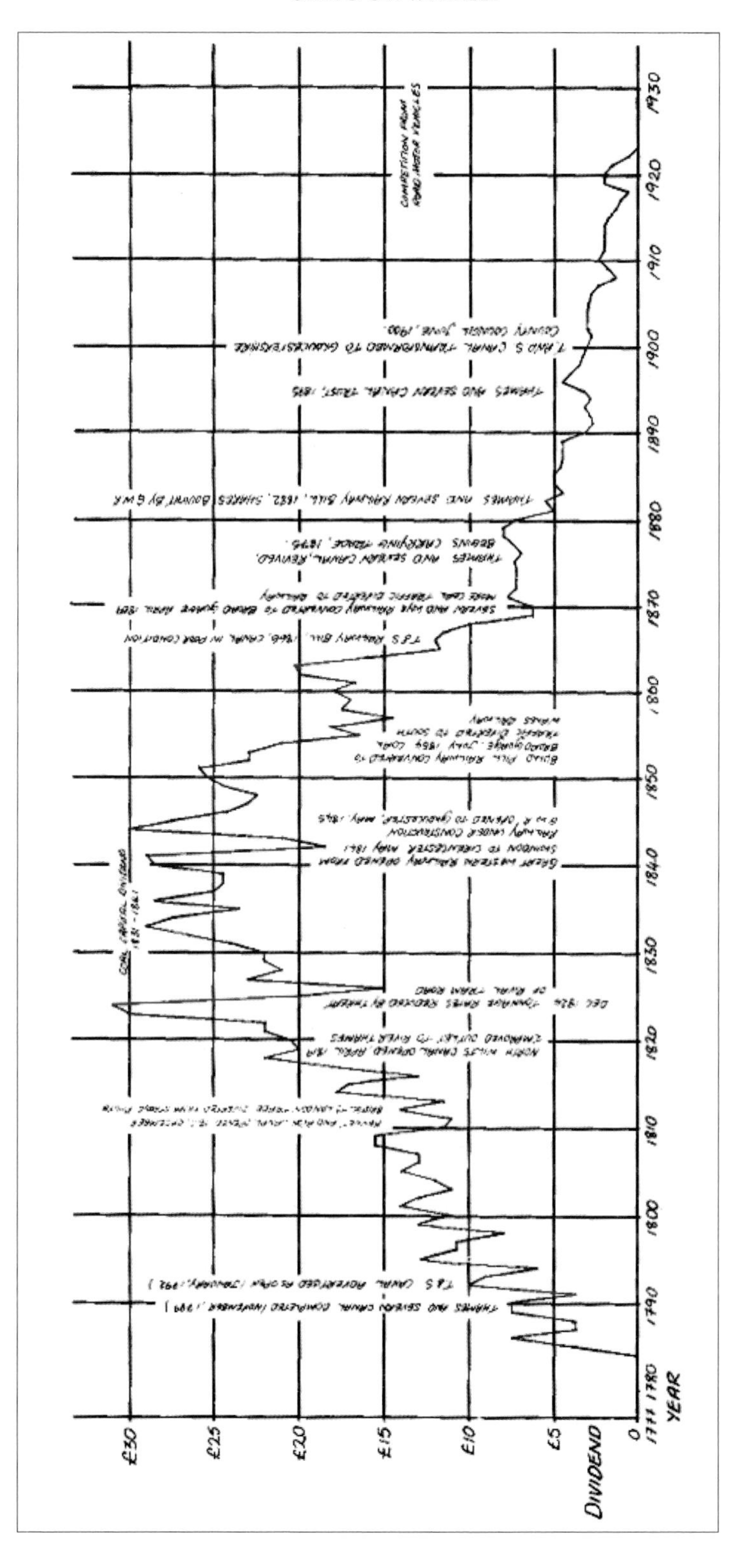

Stroudwater Navigation dividends 1785-1922.

Stroudwater Navigation, watchman's cottage and bridge, Stonehouse, in 1950.

Stroudwater Navigation, office and house of the clerk at Wallbridge, built 1795-1796, in 1950.

themselves. When even this sum had been exhausted eighteen months later, the works were carried on with the aid of private loans, subsequently repaid out of earnings, all of which were devoted to that end and to the property until April 1785. So it came about that in their Share Register the proprietors recorded with commendable exactness "that the Cost of the Undertaking must be estimated at £40,929-15-4" yet their share capital remained £20,000 divided into 200 shares whose nominal value was £100.[40]

The proprietors' frugal husbandry can only be commended, but their wisdom in not increasing the shareholding may be questioned. Their first dividend was £3 15s. paid late in 1785. The average for the next six years was only £5 12s. 6d., but later on fat dividends were paid.[41] The benefit accruing to the district was never in doubt within a very few years it was said that the navigation had reduced the cost of coal by £5,000 a year[42] — but as time passed, public opinion, forgetful of the proprietors' years of self-denial and probably ignorant that £150 had been paid up on each £100 share, commented unfavourably on the large profits and high tolls — 3s. 6d. a ton on coal, 5s. a ton on other goods, authorised by the Act of 1730 and re-enacted in 1776.

The company's financial stringency was at first so great that even the provision of essential buildings had to wait. A warehouse and cottage had been put up at the terminal basin by 1780. A cottage costing £65 for the lock-keeper of the double lock at Ryeford was built in 1784. But the navigation office and home of the clerk at Walbridge was not built until 1795-1796, and even then there had been an initial decision to reduce its scale and cost.[43] Fortunately, for it would have spoilt the appearance of the building, this decision was later rescinded, but the early years of scrimping left a less obvious mark in the niggardliness which clung to the committee and the proprietors for a very long time.

In one important respect the canal was out-of-date from the start. Brindley's canals were equipped with horse towing-paths, but there had been no horse towing on the River Severn when Hore drew up his plan for Stroud, nor when Yeoman produced his. The committee did not look ahead, and their towing-path, unfenced and with stiles instead of gates at each enclosure, could not be used by horses.[44] All towing was therefore done by men, who took a day or more to haul a barge between the Severn and Stroud. It was labour more suited to beasts than men, and so, as an enthusiastic local poetaster wrote:[45]

> . . . when they're favour'd with a western gale,
> They quit the rope and briskly hoist the sail;
> Releas'd from toil well pleas'd they march behind,
> And view the streamer shiv'ring in the wind

(the 'streamer' being the pennant flying from the mast-head, as was customary on Severn River craft). There is in fact more evidence than one would expect for vessels sailing in the narrow confines of a canal.

The company's organisation was very simple. The Act specified an honorary treasurer, a salaried clerk, and a committee of nine to thirteen members, all of whom had to be resident in Gloucestershire. No proprietor was to hold more than fifteen shares, and at the general assemblies, held twice a year, the voting of absentees by proxy was severely limited.[46] Local control of the company's affairs was therefore ensured to an unusual degree. There were, however, a few proprietors whose interests were certainly not local, and some of these we shall meet again: Christopher Chambers, for instance, merchant of 38 Mincing Lane, London; Thomas Hayes, merchant of Wolverhampton; Samuel Skey, manufacturer of Bewdley.[47] It was a sign of the broader significance of the Stroudwater Navigation, a significance which had never been forgotten. Even before the first stone had been laid at Framilode, there had been talk of forming another company to extend the line to the Thames, and an "able engineer" — probably Thomas Yeoman — had made a survey from Stroud to Cricklade, afterwards reporting optimistically that a junction of Severn and Thames was "not only . . . practicable, but easy to be effected".[48]

Allen, Earl Bathurst, Pope's patron, lived to be over ninety: Active to within a month of his death on 16 September 1775,[49] he would certainly have known that his vision of half-a-century before was at last on the threshold of achievement.

REFERENCES

1. VCH *Gloucestershire*, vol. 2, 1907, p. 160.
2. See JHC vol. 21, p. 437, evidence of Baghot and Cripps.
3. *Mr Pope's Literary Correspondence For Thirty Years; from 1704 to 1734*, vol. 1, 1735, pp. 135-6.
4. Dallaway, J., *A Scheme to make the River Stroudwater Navigable*, 1755, p. vi. JHL vo1. 23, p. 547.
5. Mavor, W., *General View of the Agriculture of Berkshire*, 1808, pp. 438-9. Thacker, F.S., *Kennet Country*, 1932, p. 317. Willan, op cit, p. 56. Hadfield, E. C. R., *Canals of Southern England*, 1955, p. 44. Willan, T. S., *Bath and the Navigation of the Avon. Proceedings of the Somersetshire Archaeological Society*, Bath Branch, 1934-8 pp. 139-140. *The Proposals for making the River Chelmer navigable*, no date, BM 816.m.8.58.
6. Cf Yeoman's statement in 1775 (JHC vol. 35, p. 469) that vessels loaded with 70 tons "could not navigate the River" without the assistance of three or four lighters.
7. JHC vol. 21, p. 437. It is stated there that the locks were to be 7½ feet wide, which is not compatible with the passage of boats of 60 tons. It

seems more probable that the figure refers to the height of each lock.

8. Act 3 Geo. II cap 13. JHC vol. 21, pp. 545-6. JHL vol. 23, pp. 547, 550, 579.
9. Dallaway as above, *The Case of the Stroud Water Navigation*, 1775, p. 2. Wright, E. C., *The Early Smeatonians, Transactions of the Newcomen Society* vol. 18, 1937-8, p. 101.
10. Patent Office 1758/720 to which a drawing is attached.
11. Annual Register 1760, pp. 142-4, letter from Josiah Tucker.
12. Ibid, pp. 144-8, letter from Ferd Stratford. See also comments in *The Case of the Stroud Water Navigation*, 1775, p. 2, and Bridge's own evidence (JHC vol. 28, p. 425) which shows that comparison had been made with locks on the River Avon which operated very slowly.
13. Act 32 Geo. II cap 47. JHC vol. 28, pp. 398-9, 424-5, 442, 455, 466, 470, 471. JHL vol. 29, pp. 446, 449, 465, 468, 480.
14. *The Case of the Opposition to the Bill for making a Navigable Canal from the River Severn up to Walbridge*, 1776, p. 3.
15. *The Case of the Stroud Water Navigation*, 1775, p. 3.
16. There is a copy of the 1753 edition in the William Salt Library, Stafford.
17. Phillips, J., *A General History of Inland Navigation*, fourth ed. 1803, p. 222.
18. *Gentleman's Magazine* 1758, vol. 28, p. 277.
19. See *The History of Inland Navigation. Particularly those of the Duke of Bridgwater, in Lancashire and Cheshire*, anonymous, 1766.
20. Priestley, J., *Historical Account of the Navigable Rivers, Canals, and Railways throughout Great Britain*, 1831, p. 670, where it is stated that the Thames & Severn Canal "was first projected by Mr. Brindley".
21. Whitworth, Richard, *The Advantages of Inland Navigation*, 1766, pp. 9-10, 51. He is quite definite that the Duke's boats would "not bear the navigating in the tideway". See also Smeaton, J., *A Review of Several Matters relative to the Forth & Clyde Navigation*, 1768, p. 13.
22. *The Case of the Stroud Water Navigation*, 1775, p. 1.
23. *The Case of the Stroud Water Navigation*, 1775. SWN committee vol. 1,
29. December 1774. Wright, E. C., *The Early Smeatonians*, pp. 101-2. Fisher, P. H., *Notes and Recollections of Stroud*, 1871, pp. 30, 136, 142, 213.
24. *The Case of the Stroud Water Navigation*, pp. 3-4. JHC vol. 35, p. 469, evidence of Thomas Yeoman. The length is that recorded in the company's archives.
25. SWN committee vol. 1, p. 12.
26. GJ 5 June 1775.
27. JHC vol. 35, p. 469, evidence of John Jepson.
28. *The Case of the Stroud Water Navigation*, p. 2. Other sources used in this paragraph are: *The Case of the Opposition*; GJ 7 August 1775; Lems, R.

A., *The Navigation to Stroud* in *Newsletter No 6* of the Gloucestershire Society for Industrial Archaeology, pp. 43-4.

29. Act 16 Geo. III cap 21. JHC vol. 35, pp. 461, 481, 483, 502, 512, 513, 553, 581, 630. JHL vol. 34, pp. 582, 586, 589, 591, 612. GJ 12 February 1776. SWN committee vol. 1, p. 39.
30. Phillips, op cit, p. 217. Although he mentions only two brick cylinders, there are in fact three. Chambers, J., *Biographical Illustrations of Worcestershire*, 1800, p. 485.
31. SWN committee vol. 1, pp. 12, 217, 220, 222. Lewis, op cit, pp. 45-6:
32. Handford, M., *The Stroudwater Canal*, 1979, pp. 316-7.
33. SWN committee vol. 1, p. 280.
34. The print is entitled "Stonehouse Church" and is from "A Collection of Proof Prints engraved for Bigland's History of Gloucestershire as well as those published in that work, as of those that were engraved & never published".
35. SWN committee vol. 1, p. 323.
36. DNB Pingo, Thomas, Lewis and John.
37. Report to the General Assembly of the Ellesmere Canal Proprietors 27 November 1805, p. 27.
38. SWN committee vol. 1, pp. 94, 108.
39. See the comments in Rudge, T., *History of the County of Gloucester*, 1803, p. xxxi, and in Rees, A., *Cyclopaedia*, *Canal*, 1819.
40. SWN Share Register, entry following p. 200, and committee vol. 1, pp. 157, 256-7. The dividends were always distributed on 200 shares.
41. SWN Share Register.
42. Anonymous, *Considerations on the Idea of uniting the Rivers Thames & Severn*, 1782, pp. 2-3.
43. SWN committee vol. 1, pp. 306, 376, vol. 2, pp. 17, 206, 213.
44. Rees, *Cyclopaedia*, *Canal*.
45. Lawrence, W., *Stroudwater, A Poem*, 1843.
46. Act 16 Geo. III cap 21.
47. SWN Share Register: Skey 1774, Chambers 1776, Hayes 1777.
48. GJ June 5 1775.
49. Kippis, A., *Biographia Britannica*, second ed. 1780, vol. 2, Allan Bathurst.

CHAPTER FOUR

THE RIVERS THAMES AND SEVERN

While routes from Stroud to the Thames were being discussed, the Gloucestershire historian, Samuel Rudder, commented that the Thames could "not be depended on for the general conveyance" of merchandise and asked "what purpose can such a junction answer, unless the navigation of that river were improved?"[1] As the two rivers, Thames and Severn had a profound influence on the development of the Thames & Severn Canal, it is essential to examine their condition, the trade which flowed along them, and the types of vessel which were using them and were soon to use the canal.[2]

The course of the Thames was, and still is, very indirect, as there is a huge horseshoe loop north of Abingdon and another east of Reading. In Rudder's day, the river below Maidenhead was in its natural state, as all the old weirs and locks had been removed. Above that town, there were three pound locks built in the seventeenth century to make navigable the reach between Abingdon and Oxford, but there were also some forty weirs and flash-locks before reaching the head of navigation at Lechlade. By impounding the water where the river ran swiftly over shelving shallows, these weirs made navigation possible for laden barges, but they also made it very difficult. Bargemen needed great skill to ride the cascade pouring through the opened flash-lock without their barge being swamped or sunk, and those struggling up depended on the help of a winch or large gangs of men or many extra horses. Meanwhile the running flash drained the reach above the weir, so that an ascending barge often had to wait hours, in time of drought even days, before it could proceed. Moreover, the locks were through weirs privately owned, provided for private purposes, and often controlled in a way that showed scant concern for the needs of river traffic.

Obviously barge-masters exaggerated the difficulties, but criticisms from less prejudiced sources refer to millers running the water as fast as possible when business was brisk, careless of the effect on the reach above; to some who never closed the sluice until a barge arrived; to owners who failed to provide either lock-keeper or operating tackle; to lock-keepers who were also innkeepers and deliberately kept the water low so that waiting bargemen would patronise the house.[3] Taken all round, navigation by flash was just what a contemporary official called it — "a very abominable Practice".[4]

There were authorities charged with maintenance of the navigation. The

Jacobean Commission established to build three pound locks below Oxford remained incorporated and very jealous of its privileges until near the end of the eighteenth century. The City of London was responsible for the river below Staines. The remainder was in the hands of a body of commissioners set up in the second half of the eighteenth century who laboured under great difficulties. The technical problems involved in making any river navigable throughout all seasons of the year are very great, and can result even now in experienced engineers expressing quite divergent opinions; but at least it is now recognised that any plan must embrace the river as a whole. In the eighteenth century, however, the piecemeal administration made any such objective view of the Thames an impossibility. Moreover, although the Jacobean Commissioners and the Thames Navigation Committee of the Corporation of London were compact bodies, the Thames Commissioners were not. Canal supporters may or may not have exaggerated when they stated that ten thousand persons were eligible to serve as members of the commission,[5] but even those who qualified were some six or seven hundred, a hopelessly unwieldy number, and in practice overall management was exercised by a committee of fifteen, with local committees administering five districts between Cricklade and Staines.[6] With so many persons eligible to act, it was possible for any commissioner with an axe to grind to pack a meeting and secure a favourable decision when his interests were at stake — and it is known that on several occasions this was done.[7]

Most of the active commissioners were concerned only with their own locality and failed to see the river as a whole. The two most influential held widely divergent views of their powers and duties. Edward Loveden of Buscot Park near Lechlade, a very active and purposive commissioner, was a vigorous supporter of new cuts and canals which would both improve navigation and shorten the distance;[8] whereas William Vanderstegen, a particularly conscientious man who probably had a more intimate knowledge of the whole river than any of his contemporaries, held the opinion that the commissioners' oath precluded them from making cuts of appreciable length and obliged them to oppose every canal intended to draw water from the Thames or divert the navigation from any riverside town.[9]

Constitutionally weak and divided in opinion as they were, it is remarkable that they achieved as much as they did. Before the last decade of the century they made a reasonably effective navigation between Maidenhead and Dorchester by building fifteen pound locks, but below Maidenhead, where there were no existing weirs, neither they nor the city authority were able to build more than one before the second decade of the nineteenth century.[10] Above Dorchester — the section of river particularly important for the success of the Thames & Severn Canal — the commissioners were never willing to do enough, as they did not consider the traffic justified the cost of making the navigation really

Sketches of West Country barges on the Thames, 1806.

effective, and this the seven pound locks built in 1790-1791 did not do.

The incidence of shoals and drought meant that barges had to be able to lie around without harm. So the early type was flat-bottomed, with sides almost vertical, square ends which sloped up like a punt's, and a rudder hung from a projecting fin. There was no deck, so the crew's shelter was a tilt or awning supported on wooden hoops astern. A simple square sail could be set on a mast which was hinged so that it could be lowered to pass under bridges. As the eighteenth century progressed, this primitive craft was improved by slightly rounded sides, more pointed ends, decks fore and aft with a cabin below for the crew, and a fore-and-aft sail which was set by a sprit or by gaff and boom. In the main, development spread slowly from the estuary to the upper reaches, so that in the early years of the nineteenth century barges with a great variety of form and rig could be seen together, amongst them western barges which conformed closely to the primitive type.[11] Improvement of the latter certainly owed something to the boat-builders of the Thames & Severn company.

About 1775, a Thames barge using Lechlade was 87 feet 6 inches long, 11 feet 7 inches wide, drew 4 feet, and was marked with a tonnage of 65, but larger barges were used on the lower reaches, and as the years passed the size of all tended to increase.[12] As they did so, lock-owners and barge-masters indulged freely in recrimination, each blaming the other for increasing costs and the deficiencies of the navigation, while traders complained of higher charges and greater delays on passage.[13]

Costs were indeed considerable. Mulcted by tolls paid as compensation to the flash-lock owners, by tolls for the pound locks as well, by tolls for bridges, ferries and towing-paths, the master of a 60-ton barge in 1791 paid £26 16s. 6d. between Lechlade and London. Then there was the crew to be paid and fed during a voyage which generally took six or seven days down river and about eleven on return, but which might be a great deal longer. In addition, the one horse which sufficed to take the barge downstream had to be supplemented by anything from four to eleven on return, and the stresses and strains on a towing-line were so great that one costing £10 or £13 seldom lasted more than three voyages. Freight charges between Lechlade and London were 20s. a ton downstream, 25s. a ton up. But it was always difficult to get full cargoes for the west, so barges often waited hopefully for long periods before setting out again from London, and as voyages were of unpredictable length, especially in time of flood or drought, the barge-master's margin of profit was probably small. No great amount of capital was necessary for a man to set up in business, for a Thames barge cost £150 to £160, yet a barge-master who invested in more than four was exceptional and most had only one or two.[14]

Whatever the shortcomings of the navigation, a flood of trade was borne down the river towards London: 5,000 tons a year from Lechlade, nearly 70,000 tons passing Maidenhead, perhaps 90,000 at Staines. Eastbound cargoes included iron, copper, tin, brass, nails, cheese; back came groceries, timber, hides, gunpowder, Newcastle coals.[15]

Widespread dissatisfaction with the conditions of the river led to an enquiry by a committee of the House of Commons in 1793. The chairman was Edward Loveden, then MP for Abingdon. Holding pronounced views on the proper treatment of the navigation and having financial interests which would benefit from an increase of trade, he was not a disinterested party, and in the end the findings of the committee were — not entirely fairly — a vindication of his policy and a condemnation of Vanderstegen's. The report stated that no 'competent engineer' had ever devised a plan for treating the whole river, nor had any systematic scheme of improvement been adopted, and that while the commissioners had been unable to agree on any method of treating the two worst districts, they had "opposed every Attempt at Improvement by Canals". Lastly, and most important, the report pointed out that it was essential the old locks and weirs should be purchased, the practice of flashing abolished, and control of the whole river vested in one authority.[16]

It was one thing for the committee to criticise and suggest, quite another for Parliament to act. The old locks and weirs were not purchased, nor was a corporate body given control of the whole river, until commercial traffic had almost ceased and the commissioners' income had dwindled away. Then, in 1866, the Thames Conservancy was established, and by them the work of making the river fully navigable was completed between 1892 and 1928.[17]

For the success of the Thames & Severn Canal that was no help at all, for the railways had long since come into their own.

The Severn was navigable for nearly 174 miles from Pool Quay below Welshpool to King Road at the head of the Bristol Channel. Still in its natural state, it was a very imperfect navigation, but the river boatmen jealously cherished its freedom from weirs and tolls. In the upper reaches navigation was hindered by extremes of drought and flood, by shoals, shelves of rock, and fords where the depth might be 18 inches or less. The lower reaches were subject to one of the greatest tides in the world, which, driving a flood of water with gathering impetus up the narrowing funnel of the river, could change the position of a sandbank in a few hours, so that even the most experienced boatman might find himself suddenly aground. As any stranded vessel with a fixed keel would be trapped helplessly in the sand until overwhelmed by the incoming tide, sea-going ships seldom went above Gatcombe and Newnham, the ports at the limit of the neaps,[18] and the boats of the river were sturdy, flat-bottomed craft, able to lie aground without harm. The higher they went, the smaller they were: 80 tons at Tewkesbury, 60 at Bewdley, 30 at Pool Quay.[19] With sides curving downwards to meet the bottom and forwards to bluff bows, they were quite unlike Thames barges. The larger ones, known as trows, were of 40 to 80 tons, with a square stern, an open hold, and short decks forward and aft. Most had a single mainmast and topmast about 80 feet tall, carrying square sails, but some had also a mizzen-mast on which a lateen sail could be set. Some 15 to 20 feet wide and 60 to 68 feet long, a trow cost about £300. The smaller vessels, known as frigates or barges, were of about 30 tons, 40 to 60 feet long, drawing about 3 feet, and having a single mast and square sail; though some had a forecastle, others were undecked with a tilt on hoops in the stern as shelter for the crew.[20] In time, the term barge ceased to denote a frigate and was used instead to describe a large shallow-draught vessel, about 120 feet long, 16 to 20 feet wide and 5 feet deep, carrying a hundred tons or more.[21]

In the middle of the eighteenth century few of these river craft ventured below Gloucester, where their freights were transhipped to coasters,[22] but

A single-masted trow under sail on the Severn above Tewkesbury, and a second with mast lowered, 1799.

later on trows and even frigates sailed the estuary in increasing numbers, some passing up the Stroudwater and Thames & Severn canals, others reaching Newnham or Bristol. It is unlikely that they went any further until near the close of the eighteenth century when the Thames & Severn company developed trows capable of voyaging to the ports of South Wales.

The boatmen were an individual race, patient, conservative and obstinate, with a notoriety for dishonest practices which surpassed that of most other river boatmen, for they were as skilled in pillaging cargoes and poaching game as in exploiting the moods of their unbridled river. In the upper reaches, they often waited weeks before a 'fresh' from the Welsh mountains set them afloat and carried them, twenty, thirty or forty together, swiftly downstream in one or two days from Coalbrookdale to Gloucester, 68 miles, where they unloaded as quickly as possible so that they might return before the water subsided.[23]

In the estuary, navigation waited not on the river but the moon, which every fortnight sent the Severn Bore racing upstream beyond Newnham, carrying vessels to Gloucester, Tewkesbury, and occasionally even Upton-on-Severn, 68½ miles above King Road, in less than half the time they took to return.[24] Aided by freshes, tides and sails, Severn craft relied far less on towing than Thames barges. For use when danger threatened or wind or current were light in the wider parts of the river, each trow had a tow-boat rowed by the crew. When towing was necessary in the narrower reaches upriver, it was done by gangs of men using towing-paths which had existed time out of mind, but were unsuitable for horses. So expenses were less and freights substantially lower on Severn than Thames: for the distance of nearly 150 miles between Shrewsbury and Bristol, charges in the mid-eighteenth century were 10s. a ton downstream and 15s. up,[25] little more than half those charged for a comparable distance on the Thames. Nevertheless, as a business proposition, carrying on the Severn had little to recommend it, as the prolonged periods of shallow water caused such delay and so limited the lading that the boatmen's profits might shrink away to nothing.[26] Few of them owned more than one or two boats: the 376 boats on the whole river above Gloucester in 1756 were divided among 210 owners,[27] and when all vessels in Gloucestershire were registered in the closing years of the eighteenth century, 35 were owned by 27 persons, of whom eight possessed two boats and the rest only one.[28]

The chief trade of the river derived from the collieries, brickworks, tileworks, pottery kilns, furnaces and forges which crowded in and around Coalbrookdale. But in the last thirty years of the century, twelve canals were built or formed to connect with the Severn, more than at any time joined either Thames or Trent, and along these came the products of the Black Country, glass from Stourbridge, pottery and coal from Staffordshire, Manchester goods, salt from Droitwich. Only at Tewkesbury was the pattern changed by the agricultural produce brought down the River Avon from the Vale of Evesham and Warwickshire.[29] The volume of trade was immense, far greater than on the Thames above London, reaching in the second quarter of the nineteenth century about half a million tons a year between Stourport and Gloucester.[30]

While the river navigation satisfied the boatmen because it was free, it was not good enough for the merchants, manufacturers and canal proprietors who depended upon it. They saw the need of weirs, pound locks, and horse towing. A horse towing-path was advocated for years, but some dozen years of the nineteenth century passed before there was a continuous one from Shrewsbury to Gloucester.[31] As for regulation of the river, the boatmen's opposition was maintained even after they and canal proprietors alike were threatened by railway promotion, so that it was only between 1843 — three years after the Birmingham & Gloucester Railway had been opened — and 1871 that six weirs and pound locks were provided between Stourport and Gloucester.[32]

On the Severn, as on the Thames, the work of making good the navigation was completed far too late.

REFERENCES

1. Rudder, S., *A New History of Gloucestershire*, 1779, p. 518.
2. For the history of the Thames Navigation, see Thacker, F. S., *The Thames Highway*, vol. 1, *General History*, 1914; vol. 2 *A History of the Locks and Weirs*, 1920.
3. There is a great deal of contemporary material. This paragraph is based on: Commissioner, *The Present State of Navigation on the Thames Considered*, second ed. 1767, p. 48; *Reports of the Engineers appointed by the Commissioners of the Navigation of the Rivers Thames and Isis*, 1791, p. 12; PP 1793 Reports 13/109, pp. 24, 88; *Report by Mr Mylne, on the present state of the Navigation of the River Thames, between Maple-Derham and Lechlade*, 1802, pp. 12, 24, 46, 56; Westall, W., and Owen, S., *Picturesque Tour of the River Thames*, 1828, pp. 3, 4; and, in GCRO, Sill's Report of 8 December 1796, p. 28.
4. PP 1793 Reports 13/109, p. 7, evidence of Charles Truss.
5. Berks & Hants Canal, pamphlet c. 1825, p. 5.
6. Act 11 Geo. III cap 45. PP 1865 Reports vol. 12, p. 611.
7. See comments in Vanderstegen, W., *The Present State of the Thames Considered*, 1794, pp. 7-9; and Berks & Hants Canal, p. 6.
8. See Mavor, op cit, p. 436.
9. Vanderstegen, op cit, pp. 26, 58. Thacker, op cit, vo1. 1, p. 159, referring to Counsel's opinion of c. 1802.
10. PP 1793 Reports 13/109, p. 28. Vanderstegen, op cit, p. 8. For the dates of building the pound locks I am indebted to the Secretary of the Thames Conservancy.
11. Pyne, W. H., *Microcosm*, 1806, vol. 1, plate 6. Pyne, W. H., *The Costume of Great Britain*, 1808, plate 35. Cooke, E. W, *Sixty Five Plates of Shipping and Craft*, 1829, plate 16. Carr, F. G. G., *Sailing Barges*, revised ed. 1951, pp. 8, 15-17, 33-5, 39-43.
12. Commissioner, op cit, 1767, pp. 41-3. Mavor, op cit, 1808, p. 431.
13. See for example: JHC vol. 21, p. 427: *The case of the Barge-Masters and others*, broadsheet undated (BM 816.m.8.49); Commissioner, op cit, pp. 1-3.
14. *Reports of the Engineers*, 1791, pp. 57, 59-60. PP 1793 Reports 13/109 pp. 6, 23-26, 97. Report of J. Sills, 8 December 1796, List of Tolls Appended (GCRO). Allnutt, Z., *Considerations on the Best Mode of Improving, the present imperfect state of the Navigation, of the River Thames from Richmond to Staines*, 1805, p. 7. Commissioner, op cit, pp. 41-6. Allnutt, Z., *A New List of Barges on the Thames Navigation*, 1812 (ms Reading Public Libraries, Treacher Collection).
15. Vanderstegen, op cit, p. 43 — the gross receipts in 1793 at 2½d. per ton at

the five upper pound locks had amounted to £264 12s. 0½d., indicating a fraction over 5,080 tons. PP 1793 Reports 13/109, pp. 11, 25. Mavor, op cit, pp. 433-4.

16. PP 1793 Reports 13/109, pp. 32-5.
17. Thacker, op cit. vol. 1, pp. 192-3, 239-242.
18. For Gatcombe, see Bigland, R., *Historical Collections relative to the County of Gloucester*, 1791, vol. 1. p. 102. For Newnham, see Rudder op cit, 1779 p. 571. GJ 28 November and 5 December 1791 reported: the enthusiastic reception of the *Elizabeth* from Oporto as it had "become a question whether such a brig . . . could navigate the Severn so high".
19. Harral, T., *Picturesque Views of the Severn*, 1824, vol. 1, p. 77.
20. For the trows and frigates of the Severn, see: *Gentlemen's Magazine* 1758, vol. 28, p. 277; Whitworth, Richard, *The Advantages of Inland Navigation*, 1766, pp. 9-10, 51; Allnutt, Z., *Useful and Correct Accounts of the Navigation, of the Rivers and Canals West of London*, undated but between 1805 and 1810, p. 16; *The Mariner's Mirror*, vol. 2, p. 201, vol. 26, p. 286, vol. 32, pp. 66-95, articles by Nance, Greenhill, Farr; Carr, op cit, pp. 160-7.
21. Allnutt, Z., *Useful and Correct Accounts* . . ., p. 16; Dupin, Baron, *The Commercial Power of Great Britain*, 1825, vol. 2, p. 333 note; *Edinburgh Encyclopaedia*, vol. 15, part 1, 1830, p. 242.
22. Gloucester had few trows but many coasters, see *Gentleman's Magazine* 1758, vol. 28, p. 277; and Willan, T. S., *The English Coasting Trade 1600-1750*, 1938, pp. 174-6, 221.
23. Randall,. J., *Broseley and its surroundings*, 1879, pp. 165-6; VCH Shropshire, vol. 2, p. 426, where Randall's description is repeated almost verbatim; Randall, J., *The Severn Valley*, 1882, p. 134; Walker, J., *River Severn, Report to the Committee of the Gloucester and Berkeley Canal Company*, 1841, p. 8.
24. PP 1847-8 Reports vo1. 31, pp. 438, 447. PP 1849 Reports vol. 27, pp. 186-8.
25. *Gentleman's Magazine* 1758, vol. 28, p. 277.
26. Randall, J., *The Severn Valley*, 1882, p. 134. Plymley, J., *General View of the Agriculture of Shropshire*, 1803, p. 286. Williams, E. J., *The River Severn as it was, is, and ought to be*, 1863, pp. 54-5.
27. *Gentleman's Magazine* 1758, vol. 28, p. 277.
28. Register of barges, trows, etc, 14 July 1795 to 24 July 1797 (GCRO). 85 vessels were registered, but 50 of these belonged to the Thames & Severn and Stroudwater companies.
29. Whitworth, Richard, *The Advantages of Inland Navigation*, pp. 41-3. *Proposed Improvement of the Severn Navigation*, broadsheet undated c. 1786. Harral, op cit, vol. 2, p. 3. Randall, J., *The Severn Valley*, pp. 419,

428. Willan, T. S., ECH 1937, vol. 8, pp. 78-9.
30. PP 1847-8 Reports vol. 31, p. 468. PP 1849 Reports vol. 27, p. 186.
31. Plymley, J., op cit, p. 314. Priestley, op cit, pp. 594-6.
32. There is much contemporary material in Gloucester City Library, MF.1.53.1-4, 7947.203, and HD.2.2-5. See also PP 1847-8 Reports vol. 31, pp. 429-475; PP 1849 Reports vol. 27, pp. 177-207; PP 1852 Accounts & Papers vol. 49, pp. 490-501; Williams, E. J., *The River Severn as it was, is, and ought to be*, 1863; Marten, H. J., *The Severn Commission, Report . . . upon the Past History, Present State, and Further Improvements*, 1892.

CHAPTER FIVE

BIRTH OF THE COMPANY

"The Company of Proprietors of the Thames and Severn Canal Navigation": the title, which followed current practice, has a panache altogether lacking in the impersonal '& Co.', and the wording reflects a fundamental difference between the subscribers to the canals and the shareholders in later enterprises; for in general, the British canal system was financed by local interests for local benefit, not by investors looking for a profitable return on their capital. This was a point which impressed Baron Dupin, the Frenchman who made a study of British public works shortly after the end of the Napoleonic wars, and who noted with surprised approval that "In England it is remarkable how many persons present themselves as subscribers to any new canal, and yet the majority are not capitalists. The proprietors of ground, of mines, or of manufactures, in the vicinity of the course of the projected canal, calculate . . . the more extended sale and the increased value of their agricultural and manufacturing produce; advantages which indeed are far superior to the profit most of them will derive from their shares in the proposed canal. This is a happy spirit of foresight and calculation."[1]

Few Acts betray their promoters' interests, but amongst the host of names it is possible to identify here and there those of landed aristocracy and gentry, doctors of medicine and divinity, clergy, mine owners, manufacturers, merchants, bankers. Even before Dupin wrote, however, it was becoming common for the last four types to show interest in canals far from their own neighbourhood and to exert an influence which was not always compatible with local needs.[2] The risk of this influence had been foreseen by Sir Richard Whitworth, who had suggested limiting the number of shares permitted to be held by any one individual and also the number of votes he might cast at the meetings.[3] But there was no way of preventing a proprietor subscribing in the names of several members of his family or subsequently acquiring additional shares through marriage settlement or bequest.

The company's affairs were directed by a managing committee (seldom then called directors) who could restrain the influence of financiers to some extent. Few went as far as the Stroudwater, which insisted that all members of the committee should be resident in Gloucestershire,[4] but many insisted that decisions should be made by majority vote regardless of the size of individual shareholdings.[5] Apart from the committee, management was sketchy: the office of chairman, so essential a feature of later company direction, had not acquired significance;[6] "the duties of manager", wrote Dupin, could not "be entrusted to any agent of the company", but should be "performed gratuitously",[7] presumably by some member of the committee; there was a treasurer who was generally, though not always, a banker; and there was a salaried clerk to register the shares and keep the minutes, who in the case of the Stroudwater had to be a full-time employee resident locally, but in that of the Thames & Severn might live at a distance and handle much other business.

Promotion of the Thames & Severn, however, did not conform to the common pattern, for true to the generations who had believed in it as a work of national importance, the promoters were not primarily local people. Some of the most prominent were Londoners, others were proprietors of four canals who wanted an outlet from the South Staffordshire coalfield to the Thames. One of these was the Stroudwater, and the others were the Staffordshire & Worcestershire, Stourbridge and Dudley canals whose only outlets at that date and also the Birmingham Canal's only outlets were southward to the Severn or northward over the Trent & Mersey Canal. These four were particularly active in the early stages of promotion, but for some time they could not agree upon a route.

The Stroudwater proprietors were naturally eager that the new canal should start from Stroud, and it was their committee who took the first practical step by commissioning a survey from Stroud to Cricklade in April 1781. It cost them £29 19s. 7d. The surveyor's report, which wisely recommended joining the Thames at Lechlade rather than Cricklade, was in their hands a fortnight later when they wrote to the Staffordshire & Worcestershire optimistically telling them "of the Practicability of a Junction with the Thames". A few months later they commissioned another and more accurate survey.[8] But the Stourbridge proprietors, probably also those of the Dudley, were less enthusiastic. For one thing, they felt that the very high tolls of the Stroudwater company would hinder the development of through traffic.[9] For another, it was obvious that a shorter line from the mining area to the Thames could readily be found, and in fact one from the Severn at Tewkesbury through Cheltenham to Lechlade was surveyed at the expense of "some gentlemen of Worcestershire"[10] who were probably proprietors of those companies. No doubt this was why a deputation from Stourbridge went on 8 November 1781 to discuss the proposed junction with the committee of the Staffordshire & Worcestershire," which, as by far the most important of the four companies and one of the most prosperous canals in the country, was in a strong position to act as arbiter between the contending factions.

Meantime a meeting in support of the Thames & Severn had been held at The King's Head in Cirencester on 17 September with Edward Loveden in the chair, when a committee was formed and a subscription list opened.[12] Naturally enough in that setting, local interests asserted themselves in favour of the Stroud-Cirencester-Lechlade route. Nevertheless, the controversy persisted, and the following year it was aired in print in a slim book entitled "*Considerations on the Idea of uniting the Rivers Thames & Severn through Cirencester with some observations on other Intended Canals*". It was published anonymously, but internal evidence suggests that it was written by a landowner,[13] certainly by someone well versed in canal management, intimately concerned in the affairs of the Staffordshire & Worcestershire company, and uncommitted to either the Stroud or Stourbridge factions.

The author showed that the chief objection to the Cirencester route had already been removed, as the Stroudwater company had undertaken to reduce their tolls on through traffic, and that being so, he claimed that the Cirencester route was in every way superior: the "experienced engineers" who had already been engaged were in favour of it; starting from Stroud, there was a shorter distance to cut and fewer locks to build; there was a better supply of water which would be obtained with less damage to existing mills; there was a better prospect of attracting trade from the Stroud valley, Bristol, South Wales and the Forest of Dean. Basing his calculations upon an estimate of £112,000, the author reckoned that a capital of £130,000 would suffice, and quoted figures indicating that there should be no difficulty in earning a dividend of 5 per cent.[14]

Even so, the Stourbridge party remained unconvinced, and, unable to reach a decision, the promoters sought expert professional advice. Of the six or seven engineers whose services were most eagerly sought during the canal era, Brindley and Yeoman were then dead, William Jessop was still little known, Thomas Telford was an obscure stone mason in Scotland, and John Rennie was a student at Edinburgh University,[15] leaving John Smeaton whose distinction had been won primarily in other spheres of engineering, and Robert Whitworth,[16] the ablest of Brindley's pupils, a man of high integrity, and a surveyor and draughtsman of great skill who had had "more Experience in [levelling navigations] than any man of his Profession".[17] Whitworth it was whom the promoters invited to meet them at Cirencester on 22 December 1781, and as the Stroudwater company had already spent a good deal on preliminary surveys, the Staffordshire & Worcestershire, Stourbridge and Dudley companies agreed to share the expense between them. It proved to be no more than eighteen guineas and Robert Whitworth had to wait at least a year before he got even that.[18]

Whitworth's commission was curiously imprecise. He had not been told what type of vessel was to use the canal, and on that depended the dimensions of the cut and the locks. Three possibilities had been considered by the earlier surveyors, Severn trows, Thames barges, and narrow canal boats, although it was well known that trows were too wide for the locks on the Thames, barges too long for those on the Stroudwater. Failing to foresee the inevitable spread of the narrow boat from the midlands to the Thames, or even the importance of carrying from Stourport to London in one bottom, the promoters had made no decision other than the vague one that cargoes would have to be transhipped between trows and barges at some unspecified point in the Stroud valley. Whitworth therefore had to make his own choice. Remarking that one size would serve as well as another in comparing the cost of the rival routes, he chose the Thames barge, and assumed reasonably enough that the point of transhipment would be at the western end of the line.[19]

The promoters had asked for his opinion, but not knowing how much they expected him to do before giving it, Whitworth assumed that they did not intend him to make "a regular Survey through the Country; but only . . . take such necessary levels" as would enable him "to form a proper Judgement of the Surveys already made". His examination of these soon revealed that the details of the Cheltenham line were so imprecise that he would certainly have to resurvey the most costly part, the summit level, of each line before he could make any fair comparison. Having much other work on hand (his services were in great demand), he was not able to visit the Cotswolds until 27 October 1782, and even then it was not until 22 December, almost a year after the meeting in Cirencester, that he completed his report.

This disclosed that several miles along the summit of the Cirencester line lay "over some bad Rocky Ground" which was "worse than [he had] ever seen any Canal cut thro' for such a continued Length", whereas the summit of the Cheltenham line lay through good ground. On the other hand, supplies of water to the summit were only about a third of those available to the Cirencester summit from the Churn and the Frome. As for the estimated cost, that of the Cheltenham line was greater, £148,316 as against £127,916, for although there would be half the length of tunnelling, there would be many more locks.

His report enabled the promoters to reach a decision at once, and thereafter they hustled proceedings to an extent which can seldom, if ever, have been equalled, for there was a plan afoot, which the author of *Considerations* hoped to scotch, for providing the Birmingham Canal with an alternative outlet to the Thames than via the Staffordshire & Worcestershire and the Severn. Meeting at The King's Head in Cirencester on 17 January 1783, they adopted the Cirencester route without more ado and decided to apply for the required land immediately, hoping that negotiations could be completed quickly enough for a Bill to be introduced in the current session of Parliament.[20] Three days later, Benjamin Grazebrook, that versatile townsman of Stroud, was visiting landowners and marking out the line.[21] Edward Eliot proved to be dissatisfied with the proposed route through his Down Ampney estate and asked Smeaton to survey an alternative; as he was

MP for Cornwall and "possessed of vast borough influence", this had to be accepted in toto, without even the limit of deviation normally allowed up to 60 yards on either side.[22] But Grazebrook's negotiations were completed with an ease that an Oxford paper described as "rather uncommon, the Land Owners, in general, . . . [being] Favourers of the Plan".[23]

Powerful opposition from a different source which had discomfited several canals and which the author of *Considerations* expected the Thames & Severn also to face, surprisingly failed to make itself heard: that from the Tyne coal trade on the ground that any canal intended to supply the Thames Valley from inland collieries would injure the coasting trade, "that great nursery of seamen for the Navy", a plausible argument which had aroused prejudice as early as the seventeenth century and continued to do so into the nineteenth.[24]

In drafting their Bill, the promoters made a mistake which afterwards cost them dear. Robert Whitworth's estimate had exceeded by nearly £16,000

ANNO VICESIMO TERTIO

Georgii III. Regis.

CAP. XXXVIII.

An Act for making and maintaining a Navigable Canal from the River *Thames*, or *Iſis*, at or near *Leachlade*, to join and communicate with the *Stroudwater* Canal at *Wallbridge*, near the Town of *Stroud*; and alſo a Collateral Cut from the ſaid Canal at or near *Siddington*, to or near the Town of *Cirenceſter*, in the Counties of *Glouceſter* and *Wilts*.

The first page of the Thames & Severn Canal Act, 1783.

that upon which the author of *Considerations* had based his calculations. Yet although Whitworth had expressly excluded the cost of wharves, warehouses, reservoirs, and compensation to mills likely to be deprived of water, the capital authorised by the Bill was no more than £130,000 as proposed in *Considerations*, leaving little over £2,000 to pay for those essential features, let alone the all-too-common experience of cost exceeding estimate. Had they decided to do so, the promoters would have had no difficulty in raising more capital, for by 3 February, £103,600 had been promised, and Londoners were said to have "so favourable an idea" of the scheme that they would have subscribed a million if necessary.[25]

On 20 February, the Bill was introduced in the Commons by Sir William Guise, MP for Gloucestershire. Helped by the parliamentary influence of several promoters, Sir Edward Littleton and Loveden especially, it passed easily through all its stages (in the Lords there was not "the least appearance of an opposition"), and obtained Royal Assent on 17 April just three months after the decisive meeting at The King's Head.[26]

Remarking that canal building often suffered because no knowledgeable proprietor would keep a watchful eye on the works, the author of *Considerations* had expressed a hope that "a small number of men of liberal minds, and good fortunes" would be found among the proprietors of the Thames & Severn who would "look after and complete [the] work perfectly and expeditiously" in the short space of six years.[27] He was not disappointed, for at least six proprietors gave to the works and affairs of the company a personal attention which cannot have been common.

Who then were the proprietors, sixty-three of whom were named in the Act?[28] There were members of the nobility and landed gentry, there were professional men, merchants, manufacturers, there was at least one banker. Many of them had other financial interests, particularly in mining and canals, which it was reasonable to expect would benefit from an increased flow of trade or access to new markets. Of the nobility, there were only two, Jacob Pleydell, second Earl of Radnor, and John, Viscount Dudley and Ward; neither was a large shareholder, and the latter, who was a colliery owner and proprietor of both the Stourbridge and Dudley canals and so probably one of the "Worcestershire gentlemen" disappointed by the rejection of the Cheltenham route, held only five shares which he soon disposed of. Of identifiable manufacturers, there was Samuel Skey of Bewdley on the Severn, whose works produced large quantities of acids and who was also in business as a canal and river carrier.[29] Prominent among the landed gentry was Edward Loveden, a very active proprietor whose influence with the Thames Commissioners was useful until the company's policy threatened the profits of his private lock at Buscot.

In the case of the Thames & Severn, however, it was not as with so many other canals, the part played by nobility, gentry and manufacturers which provoked contemporary comment — that was reserved for the role of the merchants. Phillips in his "*General History of Inland Navigation*" remarked of the Thames & Severn as he did of no other canal, that it was promoted "at the desire of several opulent private persons, chiefly merchants of London . . . who had no local interest", adding that "the connections of one mercantile house alone subscribed 23,000 *l.* and several others 10,000 *l.* each".[30] Individual subscriptions were limited to a hundred shares, but although Christopher Chambers of 38 Mincing Lane, London, and Morden, Surrey, held no more than sixty himself, Sophia and Frances Chambers of the same addresses held another eighty, and Robert Rolleston, merchant, also of 38 Mincing Lane, held eighty-five, an aggregate of £22,500 clearly belonging to Phillips's 'one mercantile house'. John and Matthew Chalié, two more of the largest subscribers, holding 158 shares between them, were wine merchants of 28-29 Mincing Lane. Nor were the merchants all Londoners, for James Perry and Thomas Hayes of Wolverhampton (100 shares) were also merchants, and so was Lowbridge Bright of Bristol (55 shares). The aggregate subscription of merchants was therefore not less than 41 per cent of the whole, but although several of them were very active proprietors, it was not as a group that their influence was exerted.

It was the shareholders in other canals who took the lead. Not the disappointed proprietors of the Dudley and Stourbridge canals. Not those of the Stroudwater, who having achieved acceptance of the Cirencester line, showed little interest: only five Stroudwater proprietors were original subscribers to the Thames & Severn and these included Chambers, Hayes and Skey whose investment in the Stroudwater may well, have been made in anticipation of extension to the Thames (five other leading Thames & Severn proprietors acquired Stroudwater shares later on). Proprietors of the Staffordshire & Worcestershire Canal, on the other hand, held between 25 per cent and 30 per cent of the Thames & Severn shares, and as the one coherent group bound by a common interest exercised very great influence. Chief amongst them was Sir Edward Littleton (100 shares), fourth and last baronet of Teddesley Park, near Stafford, well-known for his improvements of agriculture and cattle-breeding, five times MP for the county. He had been a leading promoter of the Staffordshire & Worcestershire Canal, was without doubt that "respectable baronet in Staffordshire" mentioned by a contemporary as one of the principal promoters of the Thames & Severn,[31] and was almost certainly the author of *Considerations*. I have been unable to prove his personal interest in mining, but as he lived within the confines of Cannock Chase, and as two of his name and title are recorded in the annals of the Shropshire coalfield,[32] it is unlikely that this enterprising man was not involved in Staffordshire mining certainly his heirs were, for Littleton Collieries were well-known at the close of the nineteenth century.

Other proprietors common to the two companies were Moreton Walhouse

Sir Edward Littleton, a principal promoter of the Thames & Severn Canal and first chairman of the Staffordshire & Worcestershire Canal.

of Hatherton, Littleton's brother-in-law; the merchant partners Perry and Hayes of Wolverhampton, Perry holding shares in the Birmingham Canal also; John Jesson, Perry's brother-in-law, a wealthy Wolverhampton attorney; Joseph Jones, a Wolverhampton surgeon and shareholder in the Wyrley & Essington Canal; James Stafford, a parson; and William Jones, a man of property with investments in the Birmingham, Stourbridge and Dudley canals as well.

Also linking the two companies were important family holdings which interwove then or later. There were the Stevensons : John, a mercer and founder of Stafford Old Bank,[33] was a promoter of the Trent & Mersey and Staffordshire & Worcestershire canals, as his son Thomas was of the Thames & Severn. The Lanes of Bentley and King's Bromley were another family with interests in several canals; their genealogy shows Johns and Thomases proliferated with bewildering frequency through the years of descent from

Thomas of 1531-1588 who developed coal mines around Bentley (and from Colonel John Lane well-known for his part in Charles II's escape after the Battle of Worcester), but John, a barrister who in due course inherited the family estates (on which was at least one colliery), and his younger brother Thomas, an attorney and Clerk of the Goldsmiths' Company, were promoters of the Thames & Severn, a John Lane was a proprietor of the Staffordshire & Worcestershire, and two John Lanes were shareholders in the Birmingham.

Several of this group of Staffordshire & Worcestershire proprietors were men of public spirit, serving as members of the commission which governed Wolverhampton during its transition from manorial control to corporate town (Perry frequently presided at the meetings of the commission)[34] and were already experienced in canal management: Littleton was a member of the committee of the Staffordshire & Worcestershire and subsequently became the first permanent chairman, Perry was their first treasurer, Jesson was their clerk, John Lane the elder handled their parliamentary business.

Here then was the nucleus of a group such as the author of *Considerations* was looking for to stimulate the management of the Thames & Severn. Littleton, Perry, Walhouse and Hayes, all members of the Staffordshire & Worcestershire committee, were chosen as members of the first Thames & Severn committee. Perry personally supervised the building of the line. In spite of the distance from his home, Littleton maintained close touch with affairs. Their efforts were ably seconded by Christopher Chambers, John Chalié, Edward Loveden, and, when he joined them the following year, James Black, a civil engineer. Thomas Lane handled the company's legal business. Thomas Stevenson was their first treasurer, succeeded on his death in 1787 by his brother William.

Once the Act had been obtained, Perry lost no time. He forsook his merchant's office for the Cotswold countryside, and within two months — before the proprietors had held their first general assembly, chosen a committee, or appointed any officers — he already had nearly two hundred men at work.[35] Recognising his signal service and the personal sacrifice involved, the proprietors at their first meeting voted him a hundred guineas for the part he had played before and after the passing of the Act, and the following year made him an allowance of £400 a year for as long as he should continue to superintend their affairs.

Josiah Clowes of Middlewich, Cheshire, was appointed resident engineer at £300 a year — "Surveyor and Engineer and Head-Carpenter" the minute book recorded — whose duties were to "assist Mr Whitworth the Surveyor in setting out the Navigation".[36] A man of 46 or 47, Clowes had been at work on the Trent & Mersey Canal.[37] As clerk, the proprietors appointed Joseph Grazebrook, Benjamin's son and successor as clerk to the Stroudwater, but probably finding the interests of the two companies incompatible, Joseph

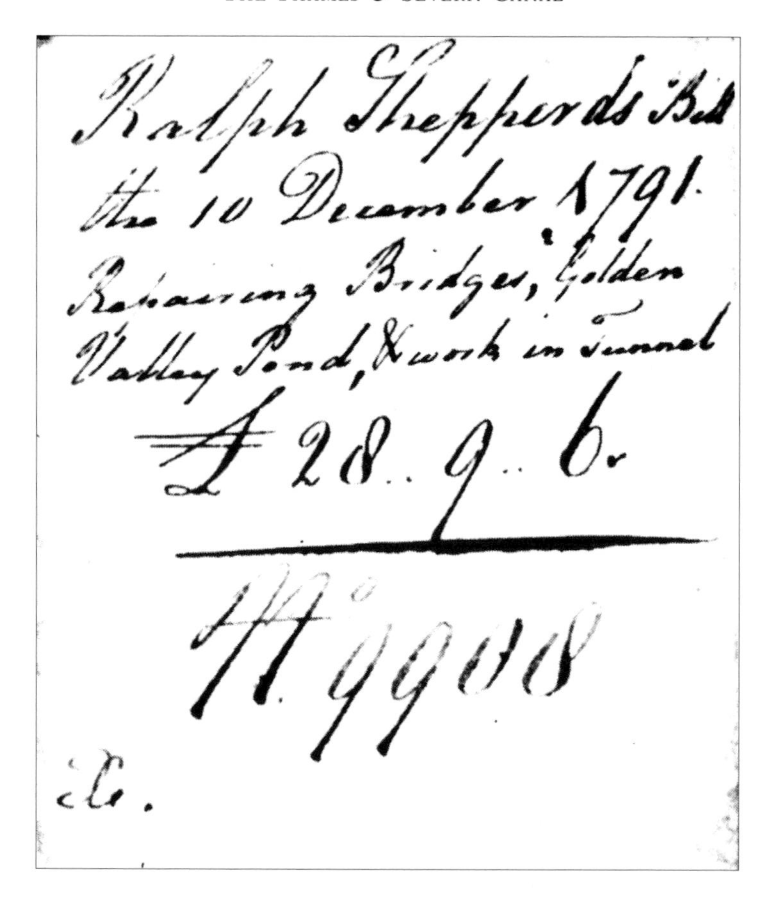
Ralph Shepperds Bill
the 10 December 1791.
Repairing Bridges, Golden
Valley Pond, & work in Tunnel
£28..9..6.
No 9908
&c.

A sample of Samuel Smith's neat docketing.

resigned after a very few months. As clerk-of-the-works to record contracts, see the engineer's designs properly executed, keep a check on materials, and pay wages and bills, the proprietors chose Samuel Smith, paying him £120 a year. With Smith's arrival from Stourbridge early in October, there began that collection of letters, memorandum books and docketed papers which testifies to his careful diligence and records with unusual detail the building of the canal. The assistant clerk-of-the-works, whose duties were to check the levels before and after excavation and measure the work of the canal cutters, was Richard Hall, a local surveyor of repute who had served the Stroudwater in a similar capacity.

In publishing their intention of establishing "an inland communication from the capital with Bristol, Gloucester, Worcester and Shrewsbury" and (through the S & W Canal) "with the manufactories in Worcestershire, Staffordshire, Warwickshire, Cheshire, Lancashire,"[38] the proprietors reiterated some of the aspirations of the seventeenth century projectors. But the days when the Thames & Severn Canal could form an important artery of commerce had long passed. During the eighteenth century the industries and population of the northern half of the kingdom had grown to an extent undreamt of. Not Bristol, but Manchester, Liverpool and Birmingham were the largest urban centres after London. Liverpool, not Bristol, was the second port. The paramount need was for communication between London and the north-west. Moreover, although a Thames-Severn canal might for a time carry to London the trade of Bristol and the manufactures of the midlands, it could not do even that for long. Inevitably Birmingham on the one hand, Bristol on the other, would establish direct communication.

Some of all this had been recognised already, for in 1768-9 the Coventry and Oxford canal companies had been formed to build a line from the Trent & Mersey to the Thames. But the Coventry soon ran out of funds and for nineteen years remained nothing but a local canal between collieries and the town. The Oxford fared little better, reaching Banbury in 1778 and being no nearer Oxford eight years later. Meantime, in spite of having thirty-two promoters in common, the two companies refused to co-operate or even join their navigations, the point of junction being a source of bitter dispute, until in 1782, on the initiative of the prosperous Trent & Mersey, agreement was reached to complete the line from Trent to Thames and to

The company's seal, from a plaster cast.

link it to the Birmingham canal system.[39] This was the plan of which the author of *Considerations* so strongly disapproved, and this was the cause of all the hurry to build the Thames & Severn Canal, for there was a very good chance that a group of determined men might complete it before the new outlet of the Birmingham Canal should be established. This aim the energetic committee achieved, for when the lines of the Birmingham, Coventry and Oxford companies were at last ready in July 1790, the Thames & Severn, authorised fifteen years after the Coventry but built in six years seven months from the passing of their Act, had already been opened.

As though someone had remembered Pope's description of the first Lord Bathurst's dream, the design of the company's seal depicted Father Thames and Madam Sabrina seated in a cavern below a wooded hill. Perhaps the active part played by London merchants in promoting the company prompted the idea of the cornucopia Father Thames was emptying into the lady's lap, but it signified little, for everyone knew that the grasping old man was less concerned with the marriage settlement he would make than the dowry he would get. TENTANDA EST VIA, the proud motto above the portal of the cavern, suggests, as the design does, that the proprietors were aware of past centuries of endeavour, and certainly shows they realised the magnitude of the task that lay ahead of them.

REFERENCES

1. Dupin, op cit, vol. 1, pp. 116-7.
2. Young, A., *General View of the Agriculture of the County of Sussex*, 1808, p. 425.
3. Whitworth, Richard, *The Advantages of Inland Navigation*, p. 20.
4. Act 16 Geo. III cap 21.
5. Dupin, op cit, vol. 1, p. 138.
6. It receives no mention in many of the early Acts.
7. Dupin, loc cit.
8. SWN committee vol. 1, pp. 346, 349, 353, 357.
9. See Stourbridge Navigation committee, 25 February 1783 (BTR).
10. Anonymous, *Considerations on the Idea of uniting the Rivers Thames & Severn through Cirencester with some observations on other Intended Canals*, 1782, p. 4 (BM 8775.f.20).
11. Staffordshire & Worcestershire Canal committee 1766-1785, p. 210 (BTR).
12. *Considerations*, p. 3, Westall & Owen, op cit, p. 10. SWN committee vol. 1, p. 353.
13. This was the opinion of F. D. Cooper, then Assistant Keeper in the Department of Printed Books, British Museum, who kindly brought the

book to my notice.

14. *Considerations*, pp. 3-5, 10-12.
15. Smiles, S., *Lives of the Engineers*, 1862, vol. 1, p. 470, vol. 2, pp. 126-8; *Gentleman's Magazine* vol. 51, p. 47, death of Thomas Yeoman in January 1781; Hughes, S., Memoir of William Jessop in Weale's *Quarterly Papers on Engineering*, 1844, p. 31, states that Jessop's "active period" was 1788-1805; Priestley, op cit, refers to Jessop's work on various rivers, canals and railways, the earliest of which is 1789. Telford, T., *Life*, pp. 15, 19; Rolt, L. T. C., *Thomas Telford*, 1958, p. 7.
16. For Robert Whitworth, see Smiles, op cit, vol. 1, p. 476; Phillips, op cit, pp. 211, 230, 544-5.
17. Wright, E. C., *The Early Smeatonians*, p. 107, quoting words spoken by John Golborne in 1777.
18. T & S Collection Letter Book, S. Smith to R. Whitworth, 16 December 1783 (GCRO).
19. Ms Report of Robert Whitworth "To the Promoters of the Navigable Communication propos'd to unite the Rivers Thames and Severn", dated 22 December 1782 (GCRO T & S Collection). Whitworth mentions the date of his first meeting with the promoters, and although this took place before "Considerations" was published, it seems fair to assume that all the arguments published therein would have been advanced by Sir Edward Littleton when the committees of the Stourbridge and Staffordshire & Worcestershire companies had met on 8 November 1781.
20. GJ 13 & 20 January 1783.
21. T & S Collection, Workmans Ledger of Debits no Credits, p. 24 (GCRO).
22. T & S Collection, Letter from Craggs Eliot to Sir Edward Littleton 15 August 1789. Act 23 Geo. III cap 38, sections 7 &.9. DNB Edward Eliot 1727-1804.
23. *Jackson's Oxford Journal* 8 February 1783.
24. *Considerations*, p. 13. Bodleian Ms Rawl A.477.1, c. 1662-4. Phillips, op cit, p. viii. Rees, *Cyclopaedia*, Canal.
25. GJ 3 February 1783. *Gentleman's Magazine* vol. 53, p. 531.
26. JHC vol. 39, pp. 148, 174, 192, 237, 240, 251, 264, 274, 297, 323, 344, 358, 386. JHL vol. 36, pp. 628, 632, 635, 639, 657. Act 23 Geo. III cap 38. GJ 14 April 1783.
27. *Considerations*, pp. 12-13.
28. Besides the names of promoters in Act 23 Geo. III cap 38, there is a list of shareholders in the Proprietors' Register 23 April 1793, and a printed list of 21 June 1808. In tracing the persons concerned and their connections, I have made use of the following sources other than those individually mentioned: Trent & Mersey Act 6 Geo. III cap 96; Staffordshire &

Worcestershire Act 6 Geo. III cap 97 and Proceedings of the Committee 1766-85 (BTR); Birmingham Canal Act 8 Geo. III cap 38; Stroudwater Act 16 Geo. III cap 21 and Register of Shareholders; Stourbridge Navigation Act 16 Geo. III cap 28 and Committee Book 1776-83 (BTR); Dudley Canal Act 16 Geo. III cap 66; Wyrley & Essington Canal List of Proprietors 12 May 1794 (BRL); Kent's *London Directory* 1783; Lowndes's *London Directory* 1787; Holden's *Triennial Directory* London 1799; Routh's *Bristol & Bath Directory* 1787; Burke's *Landed Gentry*; and most valuable, G. P. Mander's *The Wolverhampton Antiquary* 1933 vol. 1, pp. 18, 19, 20, 24, 49, 55, 58-9, 239-242, 343. I am also indebted to letters from Mr. Mander himself and the Chief Librarian of the Central Public Library, Wolverhampton.

29. Dickinson, H. W., *Transactions of the Newcomen Society* vol. 18, p. 60.
30. Phillips, op cit, p. 212.
31. Pitt, W., *A Topographical History of Staffordshire* 1817, part 1, p. 257. Pitt, W., *General View of the Agriculture of the County of Stafford*, 1794, pp. 49, 54, 104. *Gentleman's Magazine* vol. 60, p. 109.
32. Randall, Broseley, p. 67.
33. *Banker's Almanack. The Dark Horse* (Lloyds Bank Staff Magazine) vol. 11, p. 542. *Express & Star & Birmingham Evening Express* 15 June 1937.
34. Mander, G. P., loc cit.
35. GJ 16 June 1783. The first general assembly of the proprietors was held on 24 June, and the first meting of the committee on 23 July.
36. T & S Collection Proprietors Register 24 June 1783; Memorandum of Agreement between Josiah Clowes and James Perry 23 July 1784.
37. GJ 9 February 1795 recorded the death of Clowes in his 59th year. I am indebted to Charles Hadfield for knowledge of Clowes's connection with the Trent & Mersey Canal.
38. *Gentleman's Magazine* vol. 60, p. 109.
39. Acts 8 Geo. III cap 36 and 9 Geo. III cap 70. Phillips, op cit, pp. 200-206. Priestley, op cit, pp. 70-3, 178-182. Hadfield, C., *Canals of the East Midlands*, pp. 15-26.

CHAPTER SIX

BUILDING THE CANAL

"Robert Whitworth", so the company stated some years later, "was not . . . at any Time the acting Surveyor and Engineer" for the construction of their canal. In fact, the proprietors expected to get far more of his attention than he was prepared — or able — to give. He produced a map on a scale of one inch to a mile, a very real need at a date when there was no Ordnance Survey and no supply of reliable large scale maps. It was engraved on a copper plate (which has survived to this day),[1] from which impressions were available for promoters, subscribers and legislators. But although he was repeatedly asked to come and mark out the line, the only part which can certainly be attributed to him is the long nine-mile summit level where the service of a highly skilled surveyor was essential.

He marked the line of Sapperton Tunnel in August 1783, sighting by means of three beacons topped by flags, the easternmost formed with two poles rising one above the other from a tree to a height of 120 feet from the ground, and took 43 levels across the high ground, inscribing bench-marks on trees, sawn stumps and range stakes. At a later date, he spent 56 days marking out the rest of the summit; for all of which he was paid £93 7s. 6d. Elsewhere, some of the marking was done by Benjamin Grazebrook, more by Clowes and Perry who between them altered the original plan here and there, not always to advantage.

The canal surveyor's equipment was surprisingly simple considering that the accuracy required involved an allowance for the curvature of the earth. A contemporary surveyor[2] listed a levelling instrument having movement only in a horizontal plane and telescopic sights with cross-hairs in the focus of the eye-piece, tripod, graduated staff, chain, arrows with which to mark the end of the chain each time it was moved, field notebook, and the help of an assistant. Whitworth's level was probably of the type Brindley rested his hand on when his portrait was painted,[3] but although the instrument had been greatly improved since its introduction in the latter part of the seventeenth century, by 1783 it was still far from having the precision of the superb instruments developed by Jesse Ramsden and Edward Troughton in time for a later generation of canal engineers.

Once the line had been marked, the company set about getting possession of the strip of land. Negotiations were mostly handled by James Perry, who often achieved an immediate cash settlement, for landowners could choose between an annual rent for the land or outright sale based on 27 or 30 years' value of the rental: rents varied from 14s. an acre to 48s. or 50s. for a close with "several apple trees" and 60s. for a garden, but 30s. was a common figure. If direct negotiations broke down, the case was referred to independent land valuers, and if they failed to achieve a settlement, the company invoked their compulsory powers. These were two-fold: appeal to the county residents named in the Act as commissioners; and if their award should be rejected by either party, then upon the verdict of a jury which would be binding. No one was eager to invoke these powers, however, for as Perry pointed out to one recalcitrant landowner: "If you judge the Land worth more y^{n} our Survey sets forth, the Commiss^{rs} must settle the difference between us, which will put the party who is wrong to some expence and more trouble". Moreover, the commissioners were all landowners and neither as public spirited nor as impartial as they should have been: in one case, the company alleged that "the Commissioners (at least some of them) leaned very much on the side of the Plaintiff and allowed Damages . . . far beyond what Mr Capel asked or the Evidence could justify giving". Forty years later, it was the opinion of James Perry's two sons, each like his father, deeply involved in canal management, that appeals to the commissioners "almost universally" resulted in the canal proprietors being worsted. Between 1788 and 1827, the commissioners were summoned eleven times to consider six cases of dispute with landowners, but two of the meetings were adjourned for lack of a quorum, and in three cases they proved remarkably ineffective.[4]

All land conveyance was carefully recorded by a London land-surveyor, John Doyley, upon what he called a 'Rough Plan', which is in fact a delightful leather-bound volume of sixty-six pages of maps beautifully delineated in red, blue, green, light brown and black. Besides ownership of the property intersected and details of the contemporary scene, his plans reveal much interesting engineering data, particularly along the line of the tunnel.

When the land had been obtained, cutting could begin, but organisation of the labour force was a formidable problem. There were then no contractors in the modern sense[5] of a man who pledges himself to complete an undertaking in a specified time according to certain standards of quality and strength and for a price he himself has named, accepting the consequences of his own miscalculation and of damage by the elements. Such responsibility was utterly beyond the resources of working men or craftsmen, however skilled. It has already been recounted how no mason would undertake to build Ryeford Lock in ground known to be insecure, and how the architect whose structure failed found himself obliged to rebuild at his own expense. No wonder the case aroused general interest, for as a lesson in the liability of a contractor it would not have been lost on the artisans of the day. As an alternative to the early method of reliance on a man of substance who would execute the work — such a man as John Hore, proprietor, engineer and undertaker of the Kennet Navigation[6] — there evolved what might be called a gang-piecework system, under which gang leaders made individual agreements for each task they undertook. It was probably a propagation from the coalfields, in many of which it had long been the practice for groups to combine under a working leader to extract and raise coal.[7] The method was extensively adopted in canal cutting in Britain and America, on the Stockton & Darlington and other early railways, in shipyards, agriculture and steel works.[8]

This was the way almost the whole of the Thames & Severn was built, the gangs being led by such men as John Nock, Edward Edge, Thomas Cook, John Holland, Herbert Stansfield, James Jackson, all of whom were masons, Ralph Shepperd, a miner, John Pickston, William Mytton, James Adkins, who were designated 'cutters'. Several of these certainly worked on other navigations — Edge, Cook and Pickston, for example, on the Stroudwater, Nock on the Stourbridge, Edge and Nock on the River Thames.[9] Their quality varied considerably. Jackson built a bridge which, "after it was measured, fell down" and had to be rebuilt by another mason. Pickston scamped some of his work. Mytton was a careful man who wheeled "Rubble . . . into the Lane to save the Hedge from being destroyed". Holland was given a gratuity "for doing his work well". Cook became a master builder of repute. Some stayed only a short time, some six years or more — Shepperd, obviously a first-rate leader and workman, nearly ten years. The size of their gangs varied widely, surviving pay-rolls for seven gangs over a period of ten weeks showing that the smallest, a mason's, never exceeded four men, while the two largest, led by Holland and Shepperd respectively, grew steadily from 40 to 54 and 35 to 74 men. It was probably from the successful leaders of large gangs, including, as Shepperd's did, carpenters, masons, miners and labourers, that the contractor evolved, as he certainly did in the last decade of the century, such men as John Pinkerton who held, not always successfully, several contracts simultaneously,[10] and Edward Banks who rose from a modest start on the canals to become partner in the foremost firm of public works contractors in the country and to be honoured with a knighthood.[11]

Yet although the company used the gang-piecework system elsewhere, for the most difficult task of all, the driving of Sapperton Tunnel, they expected to find one man who would organise the whole labour force and accept the limited responsibility of completing it within a stipulated time and at a fixed piecework rate. Four tenders for the whole tunnel were indeed received, one from John Lowden who, although an experienced canal cutter who had been invited to complete the Stroudwater,[12] subsequently proved unreliable even as a sub-contractor in the tunnel. The tender accepted came from Charles Jones who had been strongly recommended to the committee, yet he proved quite incapable of fulfilling it so that eventually they had to take it out of his hands and let the work in lots to gangs already employed by Jones in the tunnel or by themselves elsewhere.

The gang leaders were paid 4s. 6d. per thousand bricks, 5d. a square foot for quoining, 21s. per hundred square feet for other masonry, 2¾d. or 3d. a cubic yard for cutting in ordinary soil, 3¼d. in clay, 3½d. in rubble or lock-pits, 5d. or 7d.indeep cutting, 9d. for blasting rock. There were allowances for moving spoil more than twenty yards and for tree-felling. Sixpence a square yard was paid for puddling this was the vitally important lining of the canal bed with loamy soil or clay so thoroughly mixed with water by ramming and treading that it was impervious because it could absorb no more. Tools were generally provided by the company, mason's materials delivered to the site, payments made weekly or fortnightly, money advanced if necessary. Men employed directly by the company were paid day wages: carpenters, 2s., 2s. 8d., 3s.; masons, 2s., 2s. 4d., 2s. 6d.; miners, 2s., 2s. 6d.; boatmen, 2s.; blacksmiths, 1s. 6d., 1s. 9d.; sawyers, 1s. 9d.; labourers and winders in pit sinking, 1s. 6d; labourers in open cutting, 1s. 2d.; and there were others whose work was unspecified paid anything from 6d. to 1s. 9d. A limeburner paid weekly got 9s. a week in winter 10s. 6d. in summer when daylight working hours were longer.

Some of the labourers were local men temporarily unemployed because of depression in the clothing trade.[13] Others were itinerant 'navigators' moving from one site of canal cutting to another the navvies famed for their strength, endurance and rough lawlessness. Under good leadership, such as Mytton's, these men showed respect for property, but when organisation was poor, as under Charles Jones, they soon became involved in disputes and drunken quarrels which threatened life and limb. Their effect on the locality seems, however, to have been less disruptive than usual, for the Gloucestershire press records no complaints of that 'turbulence and riot' which occurred elsewhere and established a belief that navvies were "a constant nuisance to the neighbourhood, and the terror of all other descriptions of people".[14]

A few simple mechanical aids were used. A timber railway was laid in part of Sapperton Tunnel. Water was lifted from shallow workings by men or horses turning an Archimedian screw or levering Brindley's "spoon", which had a 'flap door' of leather that opened as it was lowered into water and closed as it was raised, the water then running off through "a channel cut at the end of the spoon handle".[15] From greater depths water (or soil) was raised in buckets by men turning the jack-roll or barrel-windlass familiar to every user of a country well, or by horses winding the whim-gin of the coal pits a wooden drum turned by a horizontal beam to which the horse was yoked, so that as he trod a circular path, one end of a rope was wound in and the other unwound. But for the most part the work was done by the physical strength of the navvies with simple, sturdy tools: spade, shovel, pick-axe, mattock, grafting-tool for working the puddle, wooden rammer, wooden scoop for ladling water, barrow for removing soil, wheeling-planks with which to form the barrow-ways, horsing-blocks to raise the planks where necessary.[16] Scenes witnessed while the canal was being built probably closely resembled except in dress those photographed during reconstruction by manual labour in 1900-1907.

The men began work at Stroud, and except when the programme fell behind schedule, pushed steadily eastward so that, as far as circumstances allowed, coal from the Severn and other materials could be carried by water close to the working sites. Locally there were plentiful supplies of stone, clay and timber, and to carry these the company bought a wagon for £7, a timber-carriage for £3 7s. 6d., a tug-chain for 9s., horses six to nine years old for £7 each and a veteran of seventeen for £2 10s. Lime-kilns and brickyards (one with a capacity of 700 million bricks) were set up here and there as need arose.

The western reaches of the canal lay through what Robert Whitworth called "a populous Vale", filled with the houses and mills of Stroud, Brimscombe and Chalford. "Even the buildings of the factories are not ugly", wrote Cobbett,[17] and indeed many of them, stone-built and well-proportioned, are — or were, for they are fast disappearing — aesthetically pleasing, often with

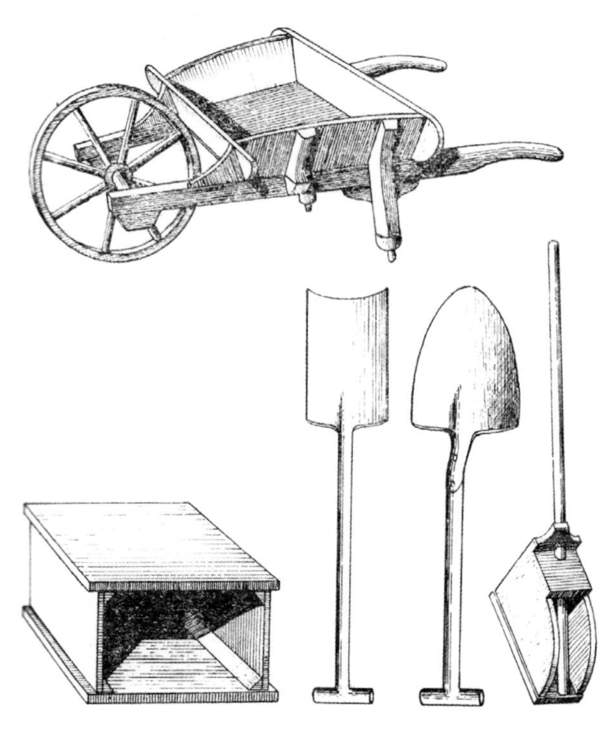

Navvies' tools: barrow, horsing block, grafting tool, shovel, and scoop, *c.*1819.

The canal approaching Chalford, with the chapel in mid-distance. Here several locks follow in quick succession, Grist Mill Lock beneath the first bridge and Ballinger's Lock beneath the next bridge up.

Red Lion Lock and Clowes's Bridge, Chalford in June 1937.
(courtesy GWR Archive ref. 12757, Paddington)

Whitehall Bridge in the upper part of the Golden Valley in 1948, "hump-backed and narrow-waisted".

Tunnellers' lodgings built at Daneway in 1784; The Bricklayer's Arms in 1947, before being renamed The Daneway Inn.

the owner's house and garden close by. Cutting was complicated by passing "very near several Gentleman's Houses, and thro' some of their Orchards and Gardens" where the work needed "more than a common neatness . . . to make it agreeable", increasing the estimated cost to £750 a mile compared with £540 in the Thames valley. The lower part of the Stroud valley is wide, the rise of its floor gradual, but in Chalford it twists abruptly, its flanks steepen and close in, and through this restricted part the channel was wide enough for only a single line of traffic. In quick succession lie three locks named from Chalford inns, one of which, Red Lion, was obviously a point of special significance as the mason Herbert Stansfield incised on the stones of the lock his own name and the date 4 December 1784, and on the keystone of the bridge crossing it "Clows Eng^{r} 1785". Now, when this point had been reached, the Black Gutter, a spectacular cluster of springs welling out of the valley flank, was available to fill the canal, and so it seems probable that on 31 January 1785, the date recorded as that when the first vessel entered the Walbridge Lock at Stroud, the canal was in fact opened as far as Chalford.

Above Chalford there were few mills, and the Golden Valley even now is much as Samuel Rudder described it in 1779, "chiefly . . . woodland, with some tillage and pasture ground, in deep hollows, and little glyns of difficult access".[18] It is not an ideal site for canal cutting: the choice lay between the bottom of the valley, liable to flood and formed of alluvial soil too porous to be safe, and valley flanks so steep that the outer bank of the channel would have to be built up. Contemporary engineering opinion was strongly against a location on such a slope, partly because of a fear that the works would be destroyed "by every sudden Rain and Thaw coming down from the Hill", but even more because of the widely accepted belief that such a built-up bank would "never be staunch".[19] The engineers of the Thames & Severn therefore chose the bottom and by so doing saddled the canal with a permanent source of weakness. A later generation, more experienced in forming earthworks, would probably have chosen the slope and risked occasional damage by storm.

As the valley floor rose more steeply, the locks became more numerous. There are twenty-eight in the seven miles above the junction with the Stroudwater Navigation, but in the upper part of the valley there are twelve in little more than 1½ miles, seven in the last half-mile leading to Daneway and the summit level, 362 feet 6 inches above ordnance datum and almost 241 feet higher than Walbridge. In the summer of 1786, the line was completed to Daneway Bridge, and there, below the summit lock, the company formed a wide basin, established a warehouse and coalyard, and built a road up the hill to Sapperton village so, that coal could be forwarded by wagon "for the supply of the country between Cirencester and Faringdon".[20] Alongside the summit lock is a house built for the company in 1784 by John Nock with the

Tarlton Bridge spanning the deep cutting east of the tunnel, as rebuilt in 1823.

local stone. It became a public house before the company sold it in 1807, and yellow-washed, with a garden sloping to the lock wall, high backed settles in the bar, and the name The Bricklayer's Arms, it long retained a delightful atmosphere characteristic of its period (as the Daneway Inn, it is rather different). A little way beyond the inn, the line curves sharply towards the steep hillside before plunging into the famous tunnel beneath Sapperton Hill and Hailey Wood.

After more than two miles under ground which slopes, as all Cotswold does, gently eastward from a striking escarpment, the line emerges in a very different setting near the village of Coates and the source of the Thames. First there is the only big earthwork on the line, the deep cutting stretching 1,223 yards from the portal to the "Level Stake" Whitworth drove in what was then Coates common field. It began at a depth of thirty feet, but had experience of earthworks been greater, it would have begun at sixty feet, reducing the length of tunnelling by some 220 yards.[21] Beyond the cutting are the bleak open spaces of the uplands where it is hard to believe the course of the canal is at no greater height than in the sheltered valley bottom at Daneway. The boundary is a dry-stone wall instead of a hedge. Loose stones lie on the surface of the fields. It is a parched and thirsty country. Even at Thames Head the meadows are bone dry except in winter. This is Whitworth's "bad Rocky ground".

There was, as he said, no doubt "but that by time and Money the ground might be moderately Water tight". But would enough time and money always be available? The intention was to give it a double lining — a puddle of stiff clay two feet thick, and above that a 'pun' or trodden mass of clay also two feet thick. Each had to be built up laboriously in thin layers, 9 or 10 inches thick, gently watered with a scoop, patiently worked with spades, firmly trodden by men wearing special puddling boots, each layer left several days before the next was grafted to it. The process was very tedious. John Pickston and his men scamped the work. Perry found they had laid only "a small puddle" beneath the pun, and wrote scathingly, bidding Pickston "as you wish to be called a workman *read and blush*". Even so there was later found plenty of evidence that the work was badly done. At a later date, this unpromising terrain would probably have been avoided. Widespread canal cutting rapidly increased knowledge of the earth's crust, and it was the canal engineer William Smith who laid the foundation of geological study. He succeeded in taking the Somerset Coal Canal across the same oolite formation by a route relatively free from leakage, and it was probably he who stated that an "equally convenient" line through better ground could have been chosen for the Thames & Severn.[22]

Beyond Thames Head Bridge and the crossing of the Foss Way, the line veers away from the valley of the Thames towards the Churn, twisting and turning along the contour until at last it strikes across a lateral valley on an embankment revetted with dry-stone walling. This leads to Smerrill where

The single-arched masonry structure of Smerrill Aqueduct over the Cirencester to Malmesbury road looking towards Kemble, *c.*1914.

the Kemble-Cirencester road was crossed by a masonry aqueduct of a single arch, a source of periodic irritation to the gentry who complained of the drips which fell on them as they drove beneath. The aqueduct has been removed, leaving on one side an interesting cross-section of the embankment and original canal bed.

The summit level extended to Cirencester, 9 3/8 miles from Daneway, but 1¼ miles short of the town the main line branched off and, through the four Siddington locks, began the descent to the Thames. Beyond this junction the scene changes again, the line quitting the uplands for the water-meadows of the Churn and Thames valleys. Both areas, however, are alike in one respect — they are sparsely populated. In the nineteen miles from the tunnel to the Thames, the canal served only six rural settlements — Siddington, South Cerney, Cerney Wick, Latton, Marston Meysey and Kempsford — and in the last three of those miles traversed lonely meadows where there was and is but a single dwelling, the isolated farmhouse at Dudgrove.

Roads and trackways crossed the canal, sometimes by timber swing-bridges, more often by brick or stone overbridges which, hump-backed and narrow-waisted, curvetted gracefully from bank to bank. Though a low aqueduct carried the canal over the Churn, there was seldom headroom for the brooks of the level water-meadows to be crossed by an arched culvert, so that they had to be led beneath either in a wooden trunk (or pipe), or through what was known as a "crooked or broken-backed culvert".[23] This was an inverted siphon, leading from a deep pit into which the stream fell, through a brick-lined passage under the canal, to a second pit on the further side where the water rose again and whence the stream resumed its course.

On the descent to the Thames, there were fifteen locks falling 129 feet in 13½ miles. Locks were a very expensive part of canal building: Robert Whitworth reckoned £80 per foot rise, accounting for 23 per cent of the cost of the Cirencester line and no less than 42 per cent of the alternative Cheltenham route. Moreover, so much more than total cost depended on them: the expense of maintenance on their shape; the consumption of water on their rise and spacing; the smooth flow of traffic on their rapidity in operation. The lock chambers of the Thames & Severn were of rectangular section, a common enough shape before engineer's discovered that walls inclined outwards and combined with the floor (or invert) in a single U-shaped structure gave greater security against the pressure of the surrounding earth.[24]

It was in the rise and spacing of their locks that the builders of the Thames & Severn made avoidable mistakes, and this was pointed out, when it was too late, by both John Smeaton and Robert Mylne. Whatever the rise adopted, it was very important — unless there were intermediate supplies of water of proved reliability — that that rise should be maintained at successive locks, so that no intervening pound should lose more water than it received or gain more than it could hold. Both in the Golden Valley and on the descent to the Thames, however, locks of differing heights were built without sufficient examination of the feeders on which reliance had been placed. As for the spacing, it was essential that the pound between any pair should be long enough to fill the chamber of the lower lock without its own depth being excessively reduced. It was considered that about a hundred yards was the minimum length, but that a pound so short needed supervision by a lock-keeper to avoid waste of water through careless closing of the paddles by boatmen. Unsupervised, 220 yards was reckoned the shortest advisable length.[25] Now, of the intermediate pounds in the Golden Valley, three were less than 100 yards long, and the top six — too many for the single lock-keeper to control adequately — were all less than 220 yards. Robert Whitworth certainly over-estimated the supplies of water, but it is probable that if he himself had marked out the whole line, he would have avoided at least some of the errors perpetrated in building the locks.

All other works on the line were dwarfed by Sapperton Tunnel, which even now has only been exceeded in length by two canal and ten railway tunnels driven in this country. In 1783, when the word 'tunnel' had only just acquired its familiar modern meaning,[26] the only comparable work was Harecastle, 2,880 yards long, 9 feet wide, 12 feet high, and as Sapperton was to be 3,817 yards long,[27] 15 feet wide, 15 feet high, Robert Whitworth was entitled to warn the proprietors that it would be "much longer and . . . wider than any that has yet been done" and that the cost was "an uncertain piece of Business". He gave them an estimate of £36,575, but although it has never been possible to calculate the actual cost, it is known that it was far more than that.

The technique of tunnelling derives directly from mining practice. Starting from the ends, if these are accessible, and working both ways from the bottom of vertical shafts sunk at intermediate points, miners first drive a low and narrow headway or driftway about the same size as a mine gallery. This they afterwards enlarge, supporting the roof and sides where necessary with timbers which remain until masons build the permanent lining of brick and stone. A straight line is always preferred because it is then less difficult to achieve the essential accuracy of alignment above and underground. Scientific alignment was not achieved until 1819, when the engineer of Frindsbury Tunnel on the Thames & Medway Canal (later used by the South Eastern Railway and known as Strood Tunnel) used an astronomer's transit telescope mounted in an observatory commanding a view of the whole of the working site. From there he first aligned the sites of the shafts. Then, as each shaft reached the correct depth, he had two lines suspended down it, each steadied by a heavy plumb-bob immersed in water. Having aligned these, he took his instrument to the bottom of the shaft and marked out the exact direction for the headway.[28] Such accuracy was not possible at Sapperton. It is probable that the intended line of the tunnel was marked by a cord stretched out across the hill, that every shaft was sited on this cord, and that as each shaft was completed two plumb-lines were hung and adjusted — on a day when there was no wind — in alignment with the cord. Maintenance of the "True Rainge" and "True Level" underground was a source of anxiety to those engaged, but although errors crept into both, they were only slight ones, and it was possible to see right through the completed tunnel by standing on one side of either portal, though not on the other. Bearing in mind the limited resources of the day, one can only marvel that the errors were not greater, and indeed there were other canal tunnels which were a good deal less accurate.

At Sapperton, twenty-five shafts were used, an unusually large number, for accepted practice then and for long after was to site the shafts at intervals of about 200 yards. They were very unevenly spaced, for one of those planned had to be abandoned unfinished, and two more than had been intended had to be sunk to meet unforeseen circumstances; the distance between adjacent shafts therefore varied from as little as 40 yards to as much as 311 yards. Each shaft was about 8 feet in diameter, a normal size, and was sunk like a well, left unlined through rock, but through soil 'steined' or lined like a well,[29] the masonry being built up on a wooden curb which sank gradually under the weight as the soil was undermined beneath it. At the head of each shaft, winding gear was set up to lower men and materials and to raise the spoil; for this, a hand-turned windlass sufficed in the early stages, but as the depth increased it was replaced by a horse-gin, one of which cost £21.

Then as now, water was generally struck somewhere while sinking or tunnelling, but it was seldom in quantities which could not be controlled by hoisting up the shaft in buckets or barrels, and this primitive method remained the standard practice a century after Sapperton Tunnel was made, pumps of any kind rarely being used, and steam pumps never unless the water was in overwhelming volume or coal was available locally.[30] Much depends on the geological strata, and although sinkers were forced to abandon one shaft at Sapperton when water accumulated above the clay line, miners tunnelling through the great oolite would have found, as did the navvies driving the neighbouring railway tunnel half-a-century later,[31] that there was natural drainage into those rock cavities known locally as 'lizens'. The complete solution of the drainage problem is, however, to drive the headway as quickly as possible from end to end; moreover to do this before opening up to the full size renders it possible to correct errors of line and level before it is too late. Even if it was intended to do this at Sapperton, untoward circumstances prevented it.

The story of the tunnel is largely the story of one man, Charles Jones, whose fecklessness not only magnified the problems and difficulties inherent in so formidable a work, but also resulted in an exceptional amount of information

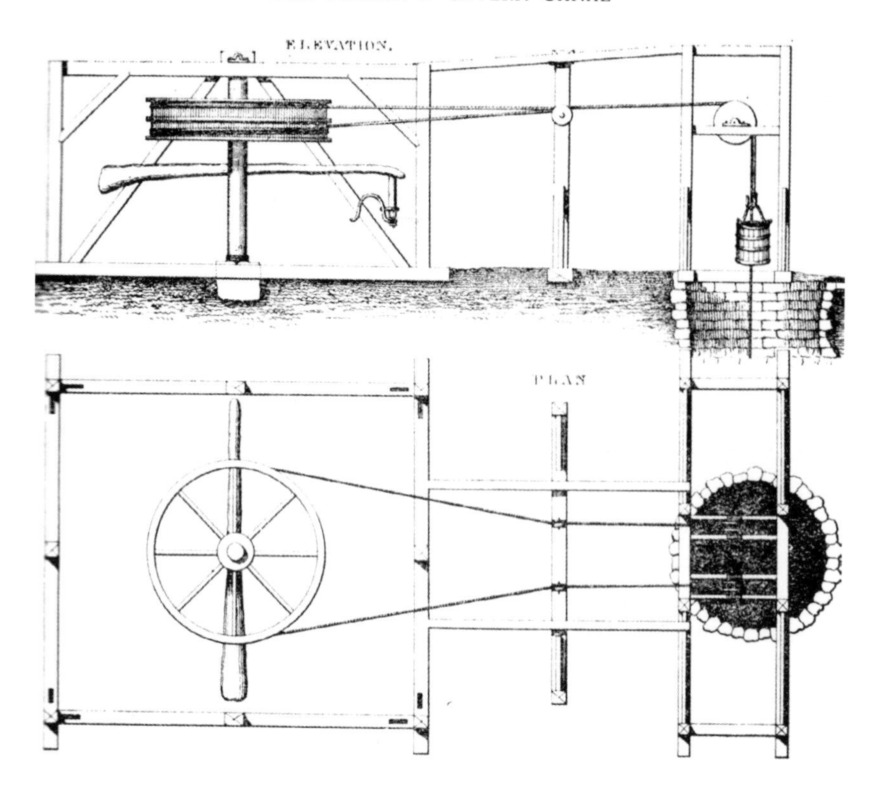

Horse gin for hoisting material up a tunnel shaft, 1826.

being recorded. At the outset, the committee were anxious enough to engage him, for he claimed to be both mason and miner, carried letters of introduction from persons known to them, and had been recommended by Josiah Clowes as one "well qualified by Experience to take the Conduct & Management" of the work. Drawn by an advertisement in the newspapers, those interested came to view the site early in October 1783, but all they were shown was two shallow shafts and a cut in the flank of the Golden Valley, which told them little about the strata likely to be encountered in the tunnel. The six tenders subsequently submitted covered shaft sinking as well as tunnelling at figures ranging from 6 to 14 guineas a yard for a bore 15 feet wide and 15 feet high, but relied on the company to provide all timber, ropes, baskets, etc. Charles Jones's price was 8 guineas, and when the committee interviewed him on 8 October, they jibbed at his figure but gave him the opportunity to revise it, whereupon he named 6 guineas, which Clowes considered too little, so the committee countered with an offer of seven. Jones then signed an agreement (it was a scrap of paper, not a legal document) accepting this figure, accepting also a vaguely worded clause that he would "follow the Directions of the . . . Engineer" if the tunnel should penetrate "other Strata or Matter than Rock", and pledging himself to complete the tunnel in four years from 1 January 1784 (Harecastle had taken eleven years).

What the committee did not know was Jones's reputation for singular ineptitude, if not downright dishonesty, in money matters. Others were better informed or more perceptive, for the two sureties he named, one a Gloucestershire man, the other a Shropshire miner, knew better than to commit themselves. He had in fact left debts unpaid in Manchester. It soon became evident that he failed to pay his men regularly. Three times he was arrested, and from gaol bombarded Samuel Smith, and even Sir Edward Littleton, with ill-spelt, pathetic notes filled with good intentions and plausible protests of his innocence, blaming his misfortunes on the machinations of ill-wishers, and pleading always for cash advances with which to satisfy his creditors' needs or his own. Nor did it help that he was a hard drinker, although that was nothing uncommon at the time, and although his drunken bouts were quite probably as much a result as a cause of his troubles, especially after the committee had added to them by treatment which was sharp and harsh.

Sharp at least was their action when, after three months' work, he reached "open Jointy Rock" which Clowes considered needed masonry lining. Taking their stand on the clause that Jones would follow the engineer's directions, the committee maintained that if they provided the bricks, Jones should enlarge the bore by 2½ feet each way and insert the lining without increase of the contract price. Maybe they were within their rights, but it was the opinion of John Taunton, the company's engineer and manager many years later, that the committee had treated Jones "rather strictly".[32] And that they themselves had some doubts was shown by their anxiety to get his signature to a supplementary agreement. So they invited him to meet them at the Swan Inn, Stroud. Wine was served, Jones was persuaded to sign, and after that, according to Samuel Smith who was present, they all sat down together and drank several glasses. A few years later Jones used this occasion as the basis of the assertion that when he signed the agreement he was "so greatly intoxicated" that he was incapable of understanding its terms. It is less unlikely than the committee's retort denying that they had ever invited him to drink with them.

A little later the committee took action which was pointlessly harsh. By this time they had learnt that Jones's sureties were not forthcoming and that no legal contract would therefore be completed. Almost certainly they had also become aware that Jones's Manchester creditors were at his heels. Quite certainly they knew that men's wages were in arrears. So they made him sign a third agreement, accepting an overseer who should disburse all moneys (including Jones's own allowance of a guinea a week), and acknowledging their right to dismiss him if he should be absent from the works for 28 days at a stretch. Ten days later he was arrested and imprisoned for debt, but he managed to get out on the twenty-seventh day. While he was inside, however, the company exercised a lien on his effects, the only security they had been able to obtain. The sale realised a paltry sum, not worth the intense resentment it provoked, for when Charles Jones and his son George next met Samuel Smith, George threatened murder. There is little doubt that father and son had been drinking, for the meeting took place at an inn, but Smith was sufficiently alarmed to obtain a warrant authorising George's arrest and was only dissuaded from exercising it by the father's intercession and the son's apologies, backed by an undertaking that George would quit the works for good.

Meantime progress in the tunnel had been niggardly. At the Sapperton end, a dispute had arisen between Jones and his sub-contractor Lowden, and what little tunnelling had been done had become so dangerous that the entrance and part of the arch collapsed in the winter of 1784. In Hailey Wood, where Jones himself was at work, there was plenty of evidence of muddle, but none of organised progress. The site of the portal had not been opened up. Nor had the lengths of headway been joined together, so that when a great volume of water was struck below the valley in the wood, it could not be run off but had to be pumped by a 'horse engine' drawing through an 8-inch pipe. Furthermore, although Jones had begun to enlarge the bore before completing the headway, he had not made it big enough to admit Clowes's 'Model or Driving Frame', with the result that the masons were unable to build the lining and so in Hailey Wood as at Sapperton, the workings had become unsafe and there was a fall of the tunnel roof and ground above it, leaving a hollow which can still be traced.

No wonder the impatient committeemen fumed impotently and sought occasion to be rid of Jones. They found it when he had another spell in prison for debt, at Lowden's suit, followed by another bout of drinking which lasted "this 2 or 3 days". And so in June 1785 they gave him three months' notice to finish off properly so much of the work as he had begun, and then to quit, at the same time notifying him that they had authorised other men to begin work at the Sapperton end. But Jones was not to be dislodged easily, and retaliated with the threat of legal proceedings. The company therefore took counsel's opinion; this was reassuring, counsel maintaining that although the proper articles of association had never been signed, reliance could be placed on the fact that such articles would have included "some guard against the Misconduct and bad behaviour of an insolvent Adventurer". As might be guessed, when the three months' notice expired, the work had not been completed. Both parties therefore agreed that arbitrators should estimate its value. The company named John Gilbert of Worsley, the Duke of Bridgwater's agent who had plenty of experience of canal works, Jones chose Hugh Henshall who had completed Harecastle Tunnel after Brindley's death. But Henshall pleaded illness and no meeting ever took place.

As late as the summer of 1788, Jones was still around when he wasn't in gaol, for he spent ten weeks there, although, mindful of his agreements and plausible as ever, he managed to persuade the authorities to let him out for a day in the middle of his term, and to drive in a chaise with a sheriff's officer to Sapperton where he descended one of the shafts. In the autumn he attended a meeting of the Basingstoke Canal proprietors and, having cajoled them into giving him the contract for their Greywell Tunnel, styled himself 'Architect of Grewell'. Four years later, by which time he was at Dudley, he or lawyers exploiting him had filed a Bill in Chancery claiming that he had completed the whole of Sapperton Tunnel within the four years stipulated. He had in fact driven about 1,487 yards,[33] not all of which had been properly finished. The company lodged an effective reply, proving they had paid out more than £14,335 on his behalf, that he was in debt to them for over £1,900, and recording that they had "found him neither a skilfull Artist, Attentive to his Business or Honorable, But Vain, Shifty, and Artfull in all his Dealings". No further proceedings followed until the company's counsel raised the case in March 1795, when Jones's Bill was dismissed with costs.[34]

Once Jones had been dislodged from at least part of the works, rapid progress was made. Five groups of fresh undertakers set to work, sinking more shafts, and for a time at least working day and night, Sundays included. They cleared about twenty yards a week, so that early in 1787 the headway had been completed throughout and the work of enlargement was going on apace.[35] The scale of the works attracted visitors, one of whom naively remarked that "as the traveller passes through the wood over the Tunnel, the noises occasioned by the rocks blown to pieces by gunpowder beneath him have a singularly romantic effect, and resemble the description of the subterraneous sounds near volcanoes".[36] Imagination boggles at the thought of men handling gunpowder by candlelight in the confined space of the headway, never more and often less than 4 feet wide and 6 feet high. Large quantities were used, for more than £3,000 is said[37] to have been spent on powder (at £3 17s. 6d. or £4 a barrel) used in making the whole canal; most of it in the tunnel, and Charles Jones's accounts alone show over £1,300 for powder and candles (at 7s. 8d. to 8s. 2d. a pound). Once the sections had been linked together; the acrid smoke and foul air at the working faces was partially dispersed by a fire kept burning at the foot of a chimney erected in several shafts. Indeed, the need of such an upcast shaft became so pressing once rock had been reached from the Sapperton entrance that at a distance of 13 chains Ralph Shepperd, who was at work there, sank a new shaft which, as it is marked by no spoil bank, can have served only for ventilation.

It was at Sapperton, where the headway was driven horizontally into the hillside, that the best progress was made, for here Shepperd was using a railway, a timber track on which ran wagons or sledge-carts with flanged wheels. It had been the property of Charles Jones, for whom a pair of "Railway Wheels" had been made in April 1785 at a cost of £4 9s. 6d., and it was described by one of the many visitors in what is perhaps the earliest published description of a contractor's temporary railway. "There is a stage or platform", he wrote, "laid for the wheels of the waggon to run on, and from a shoulder which is given to the wheel, the waggon . . . is prevented from slipping off. The wheels are one piece of solid wood". This visitor found a gang of eight men at work: three miners, two loaders, two drivers, and one man emptying the wagon.[38] Here too, in July 1787, came John Byng, later fifth Viscount Torrington,[39] who, having walked some way beside the canal in the open, "return'd to the tunnel mouth . . . whence a sledge cart issued, drawn by two horses, into which I enter'd, (seated on a cross plank,) for the *pleasure* of this inspection. Nothing cou'd be more gloomy than thus being dragg'd into the bowels of the earth, rumbling and jumbling thro mud, and over stones, with a small lighted candle in my hand, giving me a sight of the last horse, and sometimes of the arch . . . — When the last peep of daylight vanish'd, I was enveloped in thick smoke arising from the gunpowder of the miners, at whom, after passing by many labourers who work by small candles, I did at last arrive. . . . My cart being reladen with stone, I was hoisted thereon, (feeling an inward desire of return,) and had a worse journey back, as I cou'd scarcely keep my seat. — The return of warmth, and happy day light I hail'd with pleasure, having journey'd a mile of darkness."

Besides working from the Sapperton end, Ralph Shepperd drove a short length in Hailey Wood, making his share 883 yards in all. He was evidently a man of courage and determination, for he had been the first of the fresh undertakers to start work, and so, carrying a letter of authority from the company, the man chosen to test Charles Jones's reaction. In Hailey Wood he faced a far more delicate and potentially dangerous situation than at Sapperton, for his task there was to do what Jones had failed to do, open up the eastern entrance, and that he could only do in close proximity to Jones. It is certain that he was denied access to Jones's shaft, for a little to one side of the true line of the tunnel he sank his own, a shallow one marked today by a saucer-shaped depression in the paddock next the inn. There is no direct evidence who built the portals, but it seems probable that Shepperd's masons would have done so.* The design of the western was Gothic, with battlements and finials, that of the eastern, Classical. Situated in Hailey Wood, part of Earl Bathurst's famed Cirencester Park, this portal was by far the more elaborate of the two, built with huge blocks of rusticated stone, flanked by columns, topped by a pediment, and provided with two niches, two roundels, and a

* I have recently received a suggestion that Thomas Cook built the eastern portal, and possibly the western as well. I myself have seen no evidence to support this, but Cook's responsibility for both would not in the least surprise me as they have the appearance of a master mason's handiwork.

Sapperton Tunnel western portal and watchman's cottage, photographed by Humphrey Household in 1947.

Sapperton Tunnel eastern portal as sketched by Samuel Ireland in 1790, showing a Thames barge of the primitive type with punt-shaped ends. The New Inn (later Tunnel House) is on the bank behind.

space for an inscription. But the inscription was never added, and the niches seem never to have been occupied by the figures of Father Thames and Madam Sabrina for which they were undoubtedly intended.

Next to Shepperd, the most successful of the fresh undertakers was the mason, John Nock, who drove 785 yards in the middle. Then there were the partners, Jerry Lee and Thomas Barber, miners, who sank shafts and drove 370 yards of the tunnel but had to leave the lining of it to Nock's masons. Another partnership, of Pyner and George Bray, drove two short separated lengths. The fifth undertaker was Richard Jones, who sank the summit pit 244 feet deep and, driving out on either side of it, made a total of 181 yards. He was the last to finish, a matter not entirely due to the great depth at which he worked, for a note made by Samuel Smith reads: "Richd Jones . . . not in the work at 2 o'clock P.M. nor had been in the work this day all his men a Drinking except 3 men in the big Tunnel".

Most of these men secured better terms than Charles Jones had done. Richard Jones, for example, was paid 7½ guineas a yard. Sheppard's agreement promised review if it should prove "he has a bargain which bears hard upon him". Nock's contained the provision that if a "Horse Engine" should be required to draw water, as indeed proved to be the case, the company would provide it.

As the works neared their peak, a visitor was told that eight gangs were engaged, each of eight men, relieved every eight hours.[40] It is probable that there were as many as eleven working faces later, in which case between two and three hundred men would have been employed. They were drawn from many parts. John Byng noted men "from the Derbyshire and Cornish mines". Richard Jones was a 'Welchman'. Shepperd was probably from Derbyshire, where he returned when all was finished. Men from the mining areas of Somerset and Derbyshire were wedded to local girls in Sapperton parish church.[41] Many men lodged in the New Inn, built for the purpose (oddly enough by Lord Bathurst) at the eastern end of the works, where the innkeeper, John Marshall, was paid by the company to keep a "Check on the Men". The inn was later known as Tunnel House, and, until it was gutted by fire in 1952, had three storeys which remained substantially in their original condition. The upper floor, obviously used as a dormitory, stretched undivided across the full length and breadth of the building. The first floor, divided by light partitions, had once been similar. The tap-room on the ground floor was unusually large and had no doubt been the men's dining-room. Outside

The inn at Tunnel House photographed by Humphrey Household in the spring sunshine of 1947, very much in the external condition as built for tunnellers' lodgings.

there were a bowling-alley, a vast stone water-cistern, sheds and stables. The Bricklayer's Arms at Daneway no doubt served a similar purpose, though on a smaller scale; certainly John Nock lived there for a time. Charles Jones lived in the village of Coates. Married men and their dependents filled both parishes, Sapperton as well as Coates, whose registers betray the swollen populations, burials during 1784-1788 being almost double those of the preceding and following years.[42]

How many of these deaths were caused by accidents in the works, it is not possible to say. In 1786, a visitor remarked that many accidents had happened but none had been fatal.[43] He was almost certainly misinformed, for only a few months later the *Gloucester Journal* reported:[44] "In the prosecution of this work, many men have lost their lives; one man was killed

The stables and stone cistern at Tunnel House in 1949.

a few days ago by the carelessness of his companion, who suffered one of the boxes used in drawing the earth up the shafts, to fall down into the pit, which killed the person at the bottom. The men were brothers". Was it one of the Brays who was thus killed? Was that why their work was left uncompleted, as it was, so that it had to be finished by Shepperd and Nock? And was it an accident? After notice had been served on Charles Jones, he was jealous and angry and had probably uttered bloody threats himself before others retaliated by threatening, so he alleged, to kill him. In driving later tunnels there were several examples of "wanton and wicked acts" such as cutting gin ropes almost through, resulting, when the damage passed undetected, in men being severely injured or killed.[45] Accident or not, the Journal's statement that this man's death was one of many at Sapperton is confirmed by other contemporaries, one of whom recorded that "many lives were . . . sacrificed",[46] and another that men 'frequently perished'.[47] Tunnelling has always exacted its toll: twenty-six were killed when the first Woodhead Tunnel was driven; nearly a hundred, it is said, at Box; and although modern engineering practice has greatly reduced the risks, it has not been able to eliminate them, for six were killed in driving the third Woodhead Tunnel completed in 1953.[48]

By midsummer 1788, although Richard Jones was still at work, most of the structure had been finished and men were dispersing to other tasks. The shafts were closed by brick arches and filled in (though some were reopened much later for work of repair), and to comply with Earl Bathurst's wishes, the spoil banks around the four next Sapperton village were planted with those clumps of beech trees so conspicuous today. That was the summer when George III stayed some weeks convalescing in Cheltenham. Interested as he was in the spread of inland navigation, he did not fail to visit the Thames & Severn Canal. On 19 July, the King and the Royal Family spent the day with Earl Bathurst among the "enchanting woods and groves" of Cirencester Park,[49] when the eastern portal would certainly have been one of the sights. Indeed tradition asserts that the reach of canal in Coates common field, east of the deep cutting, was then named "The King's Reach" in commemoration,[50] and that the King also descended the hill from Sapperton village to see the western portal. Their Majesties, it is recorded, "expressed the most decided astonishment and commendation" at what they saw,[51] and on a later occasion when they were passing through Stroud, they "stopt some time to see the boats pass through the Lock" at Walbridge.[52]

A month later someone painted a curious water-colour sketch depicting visitors standing in the dry bed of the canal at the Sapperton portal, but although Josiah Clowes was not satisfied with the level of the tunnel floor, the impatient proprietors soon afterwards ordered the tunnel to be filled with water — strong springs flow into it below Cassey Well, below the valley in Hailey Wood, and near the western portal, as well as lesser ones elsewhere. By

Sapperton Tunnel western portal, from a water colour painted in August 1788 before the channel had been filled.

THAMES AND SEVERN CANAL

LONGITUDINAL SECTION OF SAPPERTON TUNNEL   LENGTH 3817 YARDS

HORIZONTAL SCALE - 6 INCHES TO A MILE

VERTICAL SCALE - 80 FEET TO AN INCH

N.B. From information supplied by the late Mr. Taunton, M. Inst. C. E.

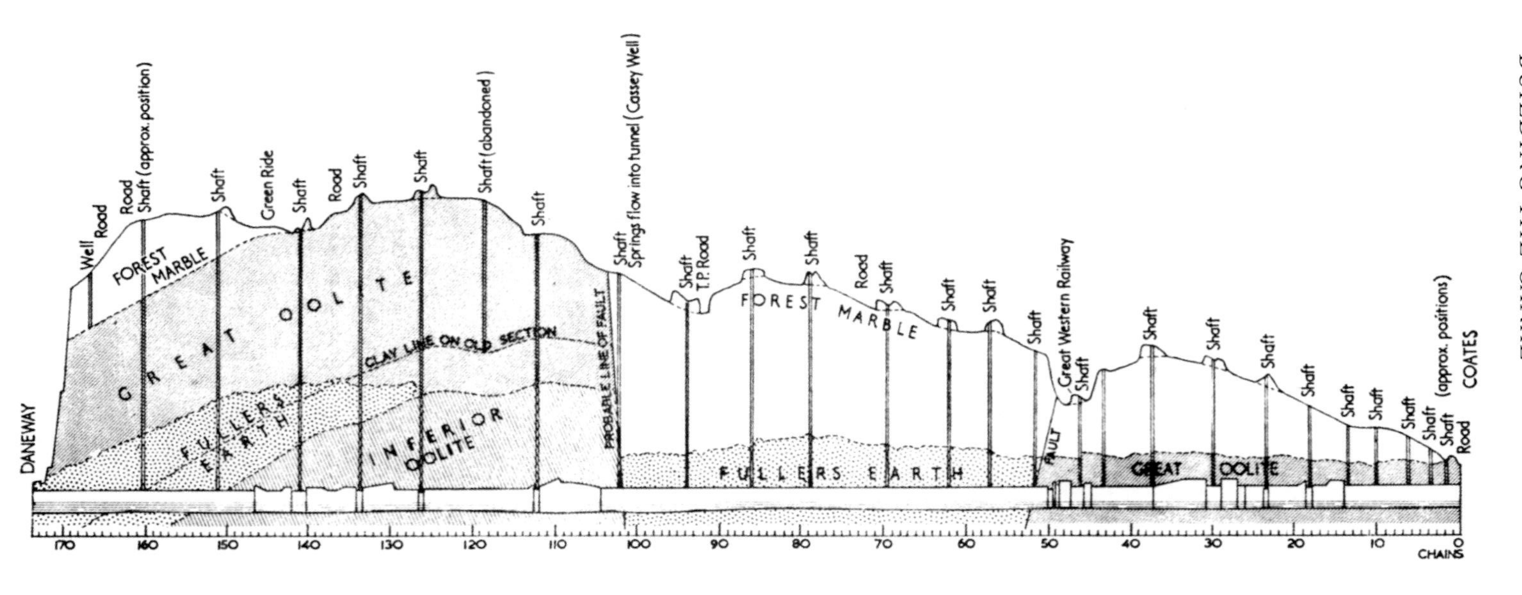

Longitudinal section of Sapperton Tunnel, from a drawing by J. H. Taunton, M. Inst.C. E.

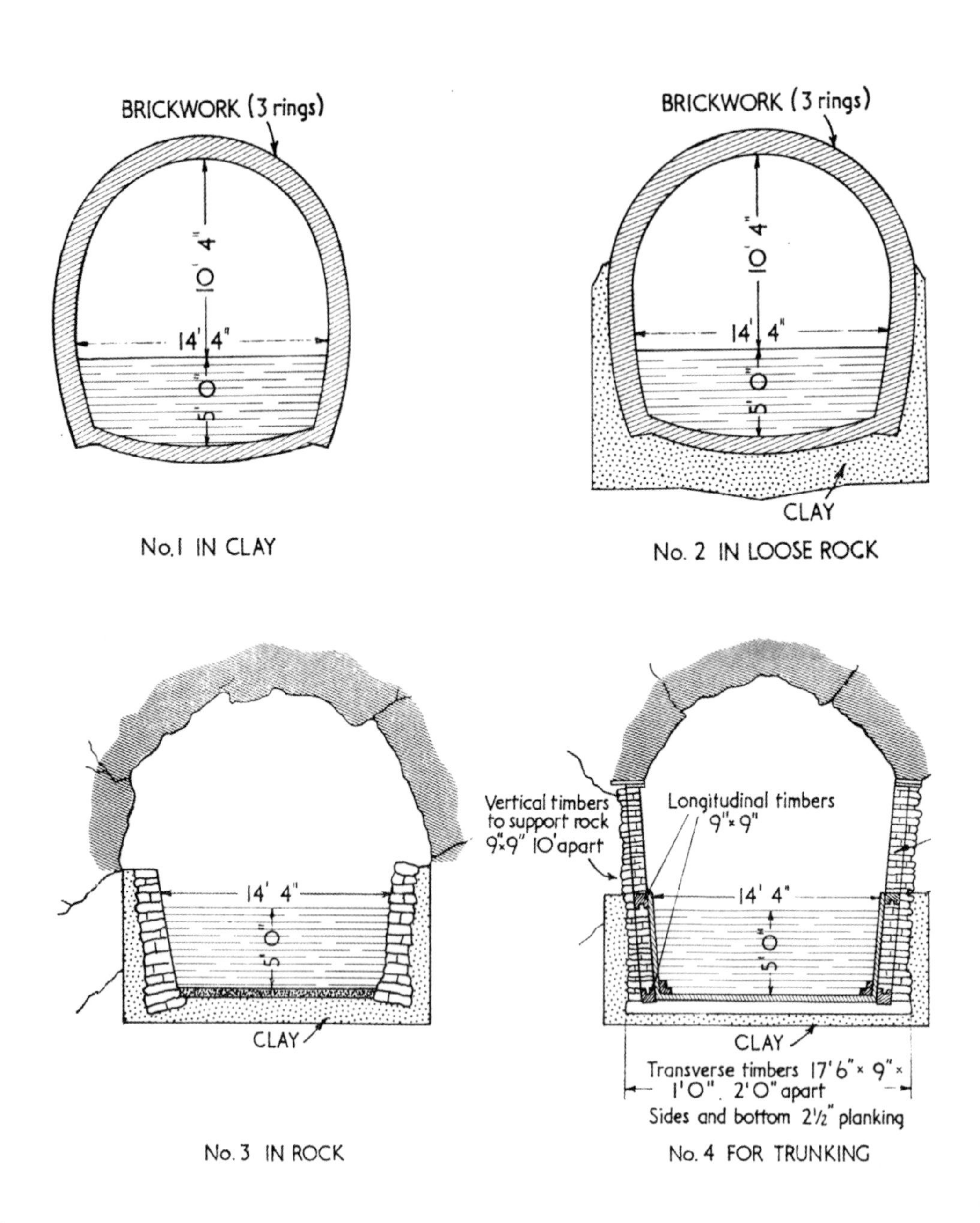

Cross-sections through the tunnel, from drawings by J. H. Taunton, M. Inst. C. E.

April 1789, it was possible — against Clowes's advice — to attempt a passage, so on the 20th, Christopher Chambers, James Perry, James Black, Josiah Clowes, and Samuel Smith set out in a boat of 30 tons. They got through, and afterwards spent the night at Cirencester, but Clowes later complained of "a Large Boat drag^{d} thro before I was ready". So the tunnel had been driven, not indeed in the four years the committee had hoped, but in little more than five, and that was excellent time compared with similar works.

The interior of the tunnel and the geological formation penetrated was described in detail by John Taunton in 1870.[53] From the Sapperton portal, the line lies through fuller's earth for about 580 yards, the whole of which had to be lined with brick or stone. Then come some 930 yards through the inferior oolite, which was left bare, except for the side walls necessary to keep the clay lining in position; "very interesting, and quite worth a visit", said Taunton of the rock here, and boatmen with whom I have talked spoke of a roof between twenty and thirty feet high in places, no doubt where the blasting charges brought down a great deal more than had been intended. Within this rock there were a great many fissures which were blocked off by oak doors, of which 160 or more were fitted into grooves cut in the rock. These served their purpose well enough at the time, but caused a great deal of trouble later when the wood decayed. A geological fault was met below Cassey Well, where the tunnel passed again into the fuller's earth for about 1,190 yards through what was known as 'The Long Arching'. This ended at a second fault under Hailey Wood, whence the final 1,120 yards penetrated the great oolite, rather more than half of which needed walling and arching. Within the tunnel, the whole channel had to be lined with clay except for some parts where sound clay was found in the beds of fuller's earth.

There was some faulty workmanship. Part of The Long Arching, for example, had been built with only one ring of brickwork instead of three. Moreover, a cavity was left above it (it is interesting that the contract for the railway tunnel specified that no such cavities should be left), so that in course of time lumps of fuller's earth fell from the roof to the crown, rolled to the sides and distorted the springing of the arch. Leakage was so severe that in 1790 the tunnel was closed for three months while men filled cracks with gravel, caulked holes with hay, and renewed parts of the clay lining. Even so, leakage persisted, and later in the same year the proprietors asked the advice of the distinguished architect and engineer, Robert Mylne, who traced the trouble to the springs which, although giving water after rain, sucked it out in time of drought. His remedy was to build a trough of planks "well caulked and pitched like the sides of a ship", supporting the base on stout cross-timbers, reinforcing the sides (heavily pounded by the passing barges) with masonry, and carrying the outflow of the springs through oaken trunks to discharge above the water-line. This was done to the parts affected, probably in 1791 when the tunnel was again laid dry for some weeks.

Nevertheless, in 1820 Baron Dupin described Sapperton Tunnel as "one of the finest specimens" of its kind,[54] and fifty years later Taunton remarked that "every expedient was resorted to in order to prevent leakage, and the engineering seems to have been of a perfect character", adding that the only improvement he could suggest lay in the extensive use which had since been made of concrete.[55]

The major cause of the defects which developed through the years was one altogether independent of the workmanship and probable wholly unsuspected by the early engineers. Many times parts of the bottom rose upwards, lengths of wall bulged inwards, and, less often, sections of roof fell in. Most of the sites where these defects developed were within the beds of fuller's earth, a substance which has the property of absorbing water and swelling as it does so. Between the masonry and the fuller's earth was the clay lining, but all clay is prone to dry and crack, and the frequent occasions when the canal was short of water or drained for examination and repair provided the opportunity for the lining of the tunnel to do so. Through such cracks, water could percolate to the fuller's earth, causing it to swell and exert pressure on the masonry. No part was more likely to suffer this pressure than the invert, or arched floor below the water-line, and indeed distortion of the invert was noted on five occasions between 1796 and 1894. On the first of these, a length of 782 yards was taken up, most of which was restored completely and reinforced with masonry struts, but the remainder was only replaced on either side of a central gap two or three feet wide. On a later occasion, in 1879, the invert of The Long Arching was utterly removed over a distance of 270 yards, and stout struts of green beech were inserted in its place. Fifteen years later, many of these struts had 'disappeared', and one surmises that they had succumbed to pressure as have other timbers whose bent and broken fragments rear fantastically from the slimy bottom at this day.[56]

Less frequent, but no less remarkable, was the effect upon the side walls. Whatever deviation there may have been from the intended width of 15 feet when the tunnel was built, in 1791 a boat 12 feet 2 inches wide passed through, but some time in the second half of the nineteenth century it was recorded that the bore was no more than 11 feet 8 inches wide at one point and 11 feet 4½ inches at another. The latter, colloquially known amongst boatmen as the 'hang up', was 160 chains from the eastern portal and a point where trouble developed on at least five separate occasions in 1809, for example, when men beginning to dismantle distorted arching were startled to find that "it came down almost of itself for seven yards together". Nowhere in the records, however, is there such dramatic evidence of the pressure exerted on arch, wall, and invert as there is in an undated and undesignated sketch of about 1879 which shows appalling distortion of all three and bears

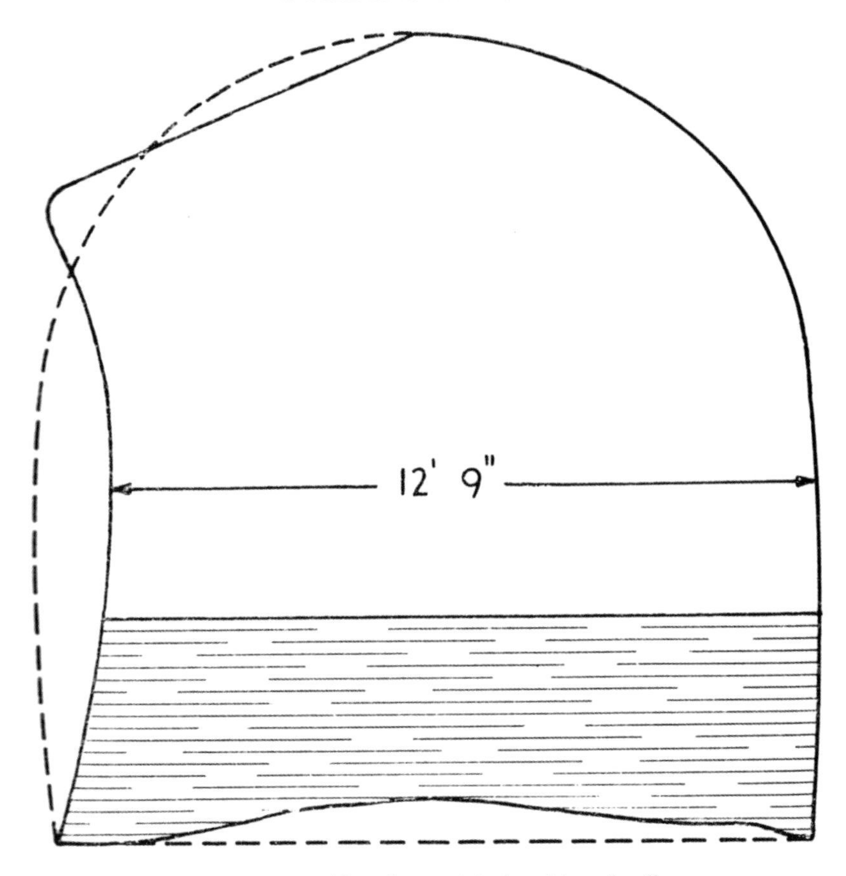

N.B. Dotted line skews original position of wall.

Diagram showing the distortion of the tunnel wall, arch and invert in the fuller's earth.

a note recording that this extended no less than 68 feet.

In most other tunnels the story is the same: formidable difficulties unexpectedly encountered; a distressing death roll; a cost that far exceeded the estimate; problems of maintenance growing ever more acute and expensive with age. All engineers disliked tunnelling: John Sutcliffe, an experienced engineer who wrote a treatise on canals in 1816, said, "I have given no instructions for executing a tunnel, because I wish another never may be made";[57] the great railway engineer Joseph Locke would have "recourse to them only when no other means could attain his object".[58]

Before the tunnel had been finished, so much work had been done along the eastern reaches that on 22 April, only two days after the first passage of the tunnel, it was possible for four barges loaded with coal to reach Cirencester, where "some thousands of spectators covered the banks of the canal, with applauding and joyful faces on so novel a sight as a river being brought, and a port being formed on the High-Woulds of Gloucestershire".[59] "How welcome", wrote another contemporary,[60] "such a sight must be in a country where fuel has been hitherto not only dear but scarce, may easily be imagined". In fact the cost of coal in the town was immediately reduced from 24s. to 18s. a ton.[61] Six days later, other barges carried coal to Kempsford.

But although before the end of April 1789, the canal was in use to a point only 3¼ miles from Inglesham, part at least of this final section had not even been marked out on the ground, for where to join the Thames had become the subject of debate. As the Thames Commissioners had so far done nothing to improve the river above Abingdon, and many believed this part of the Thames could never "be made a safe, certain and reasonable navigation", a group of promoters, including some proprietors of the Thames & Severn, had tried several years earlier to form an independent company to build a canal from a junction with the Thames & Severn at Dudgrove to the river at Abmgdon.[62] They had, however, been warned by the ancient and arrogant Jacobean Commission, then enjoying its last few years of independence, "that if no Provision was made to the Univʸ & City [of Oxford] as a Compensation for Loss of Trade the Univʸ & City wou'd oppose the Canal,"[63] a warning the canal promoters apparently ignored, for when their Bill was introduced in April 1785, it met with strong opposition, not only from university and city, and was defeated.[64]

Deprived of this solution, and time was to prove that avoidance of the river above Abingdon was the only satisfactory one, the proprietors of the Thames & Severn examined various ways of improving navigation between Inglesham and Buscot, where the state of the river was at its worst. To extend their canal was beyond their means, so they tried to persuade the Thames Commissioners to make a new cut instead of building pound locks on the river. In this they were frustrated by the rivalry between Edward Loveden and William Vanderstegen, those two prominent commissioners who held conflicting opinions on treatment of the navigation. At a meeting between members of the canal committee and the commissioners on 18 April (just two days before the first passage through the tunnel), it is probable that Loveden was present, certain that Vanderstegen was not, for the commissioners adopted Loveden's policy, acknowledging that a new cut would be preferable to trying to improve the river. Nevertheless, they refused to commit themselves irrevocably, and, after passing a vacuous resolution "that one of the two plans . . . should be adopted", adjourned for three weeks.[65] When the meeting was resumed, Vanderstegen was in the chair, and the resolution then adopted, that the commissioners had "no power . . . to make the cut or canal now proposed", embodied his view of the commission's limited authority. As this resolution was followed by another determining to improve the river without specifying how, nothing whatever was achieved.[66]

After that, there was only one thing the canal proprietors could do: follow their parliamentary line to Inglesham and hope the commissioners would act in time. So on 7 May they gave orders for the line to be marked out. In this last stretch lies the curious double lock at Dudgrove, where the upper chamber has a fall of 9 feet and is well-built with neat ashlar, whereas the lower chamber has a fall of only 2 feet 6 inches and is roughly built with unmortared stones. It has the appearance of an afterthought, as though the upper chamber had been built while the company were still preoccupied with alternative plans for junction with the river, and an increase in height had become necessary when those plans had been abandoned.

The bridge crossing the tail of Inglesham Lock is inscribed with the date 14 November 1789. Five days later, the first boat passed through it to the river, the air ringing with the cheers of the assembled crowds and the roar of twelve cannon from Buscot Park, the home of Edward Loveden. Later, the day was celebrated by dinner at five of the Lechlade inns, followed by a bonfire and a ba11.[67] The ceremony was described in the national as well as the local press, *The Times,* for example, referring to the canal as "the grandest object ever attained by inland navigation", and the *Bath Chronicle* printing a remarkable letter written by a spectator whose imagination had surely been

Dudgrove Double Lock in 1814.

fired by something more potent than mere sight:[68]

Friday, Nov 20th, 1789.

Sir

Yesterday a marriage took place between Madam Sabrina, a lady of Cambrian extraction, and mistress of very extensive property in Montgomeryshire, (where she was born) and counties of Salop, Stafford, Worcester, and Gloucester, and Mr. Thames, commonly called Father Thames, a native of Gloucestershire, now a merchant trading from the city of London to all parts of the known world. The ceremony took place at Lechlade, by special licence, in the presence of hundreds of admiring spectators, with myself, who signed as witnesses; — from whence the happy pair went to breakfast at Oxford; dine at London, and consummate at Gravesend; whence the venerable Neptune, his whole train of inferior deities and nymphs, with his wife Venus and her train, are to fling the stocking. An union which presages many happy consequences, and a numerous offspring. — I mention the lady's name, as the tendre came from her, after many struggles with her modesty and Cambrian aversion to a Saxon spouse.

A Traveller.

So the work had been done in 6½ years from the passing of the Act, and only five months more than the author of *Considerations* had forecast. It was a remarkable achievement — few canals were built more quickly.

There was further celebration in Lechlade on 29 December when the first cargo of Staffordshire coal was landed and "there was open housekeeping at the New Inn, with plenty of punch, wine, and other liquors, for men, women, and children, at the sole expence of Mr Wells", the wharfinger.[69] As in Cirencester, so in Lechlade there was good reason for satisfaction when the price of coal came down, from 30s. or 36s. a ton to 22s. or 23s.[70] A month later, the *Gloucester Journal* recorded the arrival in Lechlade of "two cargoes of fine Droitwich white salt",[71] and in April the despatch of fifty hogsheads of cider "and other heavy articles" from Gloucester to London via Stroud.[72] But the interest of the local press in these shipments only serves to emphasise their isolation. Certainly a trade in coal grew steadily, distribution covering a widening area above Stroud, but neither canal nor river navigation was yet in a state to encourage through traffic, and in fact it was not until January 1792 that the canal company inserted in Bristol, Gloucester, Birmingham and Reading newspapers an advertisement that the line was open.

Even then the state of the river was most unsatisfactory, although a lot of work had been done. In May 1789 the canal committee had notified the Thames Commissioners that the proprietors 'expected and required' the river navigation to be in a fit state by the time the canal should be opened. The commissioners had dallied far too long for that to be possible, but they did then promptly call in William Jessop, who recommended, inter alia, the building of eight pound locks and no less than fifty-five timber bridges for the towing-path.[73] Before the end of 1791, seven of these locks had been built, three to the order of the commissioners' Lechlade committee under the direction of Josiah Clowes, the others for the Oxford committee (in which the Jacobean Commission had recently been merged) by Daniel Harris, "Gaol Keeper at Oxford, . . . who . . . had never been employed in such Works before".[74] It is not surprising that within a very few months these four locks showed signs of collapse. As by then it was clear that nothing like enough had been done, the commissioners asked Robert Mylne's advice. Mylne's reports show a liking for extravagant language, but when he told the commissioners that unless eight more pound locks should be built above Abingdon, "the Navigation, across the Island, from the Severn to the Thames", would remain "totally barred and locked up", he did not exaggerate unduly.[75] Nevertheless, for the sake of an unknown volume of traffic from the Thames & Severn Canal, the commissioners were not prepared to sanction the expense.

In the meantime the canal proprietors also had been faced with many problems: criticism of their locks, an acute shortage of water, the discovery that a mistake had been made in levelling the summit, leakage in the tunnel and probably also in the "bad Rocky ground". Though Robert Whitworth was at least partly responsible for these deficiencies, it was James Perry and Josiah Clowes who incurred the proprietors' displeasure. When Perry tendered his resignation as salaried superintendent of the company's affairs in October 1790, it was accepted without even formal expression of gratitude or regret. Undoubtedly he was responsible for some mistakes, but up and down the country the professional engineers also made plenty, and it was unkind of Robert Mylne (in a pique after surveying another line of canal to Cirencester and subsequently learning that Whitworth had advised against any more canal cutting through the Cotswolds) to imply, as he did, that the "erroneous System and subsequent fatal Deficiencies" of the Thames & Severn arose from "Proprietors and busy Men becoming Engineers, self-made, and handling Subjects of the most dangerous Kind".[76] It was, after all, to Perry more than anyone else that credit was due for the rapid building of the Thames & Severn. His energy, his "trouble and attendance in Superintending the affairs of the Company", are abundantly manifest in the records as he appears now here, now there, negotiating with landowners, appointing employees, buying boats, organising trade. And it was Perry, and Perry alone, who was briefed to appear as the company's witness had Charles Jones pressed his suit in Chancery.

Clowes got the blame for the mistake in levelling the summit, although Whitworth was possibly responsible in the first place and the assistant clerk-of-the-works ought to have detected it when checking the levels after the cut had been made. There was some acrimonious correspondence about its extent, the committee contending it was 9 inches and Clowes maintaining that it was only 4½, but after tempers had cooled the matter seems to have been settled quite simply by raising the sill of the overflow weir at Siddington by 3 inches.

Nevertheless, this is one of several indications that, although no engineer could possibly be aware of all that went on while a canal was building, Josiah Clowes's supervision lacked effectiveness. Still, whatever his failings, Clowes gained from his work on the Thames & Severn as resident engineer a reputation which admitted him to the ranks of the principal engineers. He surveyed and built the Herefordshire & Gloucestershire Canal, he worked on the Gloucester & Berkeley and Huddersfield, and he drove the Lapal Tunnel of the Dudley Canal, before his final appointment as engineer of the Shrewsbury Canal, where, after his death early in 1795, he was succeeded by Thomas Telford.[77]

Although the proprietors were preoccupied with the shortcomings of their canal, the public read only the enthusiastic published tributes to an "elaborate and stupendous work of art". Between 1792 and 1828, four books descriptive of the River Thames included not only more or less extended reference to the canal, but also illustrations which are now of great interest. The most informative were Samuel Ireland's, based on sketches he had made in the summer of 1790.[78] The most artistic were the views of Thames Head and Inglesham by Joseph Farington R.A.[79] The fullest description was by William Bernard Cooke,[80] who showed that venturesome visitors were then encouraged to go through the tunnel in a boat kept specially for the purpose at Coates. Amongst these visitors, it seems, were "artists employed as scene painters to the theatres" who drew inspiration from the unlined rocks of the interior, their "variety of irregular forms, partially revealed by lights".[81]

REFERENCES

1. It is in the Gloucestershire County Record Office.
2. Leach, E., *A Treatise of Universal Inland Navigations*, c.1779-1790, pp. 126-134.
3. Mezzotint dated 1773 by R. Dunkarton (BM).
4. Minute Book of the Commissioners appointed under the Act of 1783 (GCRO).
5. NED. The word contractor was rarely used in its modern sense until the closing years of the eighteenth century.
6. Kennet Navigation Minute Book No 1 (BTR).
7. Ashton, T. S., & Sykes, J., *The Coal Industry of the Eighteenth Century*, 1929, pp. 100-113.
8. Strickland, W., *Reports on Canals, Railways, Roads, and other Subjects*, made to The Pennsylvania Society for the Promotion — of Internal Improvement, 1826 (BM 745.c.17), p. 3. Clapham, J. H. , *The Early Railway Age*, 1926, pp. 177-8, 405, 408-9, 469.
9. SWN committee vo1. 1 & 2. Stourbridge Navigation Minute Book of the Committee 11 July 1780 and 31 July 1781 (BTR). Thacker, vo1. 2, pp. 37, 43, 91. pp. 1793 Reports 13/109, p. 105.

10. Hadfield, C., *British Canals*, p. 39, Hughes, *Memoir of William Jessop*, p. 7, Patterson, *The Making of the Leicestershire Canals, Leicestershire Archaeological Society* Vol. 27, p. 19.
11. Dickinson, H. W., *Jolliffe and Banks, Transactions of the Newcomen Society* Vol. 12, 1931-2, pp. 1-8. Tharby, W. G., *The Life Story of Sir Edward Banks in The Bullet of the Merstham Community Association* 1955.
12. SWN committee Vol. 1, p. 217.
13. GJ 19 September 1785.
14. Young, A., *General View of the Agriculture of the County of Sussex*, 1804, p. 424.
15. Anonymous, *The History of Inland Navigations* . . . , 1766, p. 41.
16. Rees, *Cyclopaedia*, Plates Vol. 2, Canals Plate 7.
17. Cobbett, W. , *Rural Rides*, Everyman's Library vol. 2, p. 104.
18. Rudder, op cit, p. 292.
19. Vallancey, op cit, p. 124.
20. GJ 22 January 1787 and see also 29 January.
21. Simms, F. W., and Clark, D. K., *Practical Tunnelling* 1896, p. 3.
22. Rees, *Cyclopaedia*, Canal, entry for Thames & Severn. John Farey, the author of the article, was given "valuable hints and information" by two engineers, one of whom was William Smith. See also Phillips, John, *Memoirs of William Smith*, 1844, pp. 5, 68 — Smith knew the Thames & Severn.
23. Rees loc cit.
24. Ibid. *Edinburgh Encyclopaedia* Vol.15 part 1, 1830, pp. 303-4. Strickland, op cit; pp. 8-9.
25. Sutcliffe, J., *A Treatise on Canals and Reservoirs*, 1816, pp. 184-8. Rees, op cit. Stevenson, D., *The Principles and Practice of Canal and River Engineering*, third ed. 1886, p. 15.
26. NED quotes 1782 as the earliest literary example of Tunnel in the sense of a subterranean passage forming part of a line of communication.
27. Robert Whitworth estimated the length as 3,850 yards. The canal company's Fact Book (GCRO) recorded it as 3,808. J. H. Taunton, M. Inst. C.E., engineer and manager from 1852 to 1885, gave the length as 3,817 yards in his paper read to The Cotteswold Naturalists' Field Club (*Proceedings* Vol. 5, p. 265) 1870. Taunton's figure is likely to be the most accurate. Differences in the lengths of railway tunnels as recorded at one time and another are not uncommon.
28. Simms, F. W., *Public Works of Great Britain* 1838, division 2, pp. 2, 4.
29. NED gives steen, and examples of stein, stean and steyn.
30. Steam pumps were used in canal tunnels at Harecastle (Trent & Mersey), Standedge (Huddersfield), and Oxenhall (Hereford & Gloucestershire), also in the railway tunnels at Kilsby and Box, but for the general practice see Simms & Clark, op cit, pp. 20-22, 54-5.

31. See *Report of the tenth half yearly meeting of the Cheltenham & Great Western Union Railway* 28 October 1841.
32. Taunton, op cit, p. 262.
33. There was, indeed, a good deal of doubt about what length Jones had actually completed. Two sets of measurements recording the work of the various undertakers exist, neither completely accurate, but a close approximation. In reply to the Bill in Chancery, the company allowed Jones 1,517 yards 20 inches, a figure which is demonstrably incorrect.
34. Jones v. Corporation of Proprietors of the Thames & Severn Canal Navigation, Hilary Term, 21 March 1795 (PRO C.33/488, p. 285d).
35. See descriptions in Moreau, S., *A Tour to Cheltenham Spa*, third ed. 1788, pp. 143-5; and GJ 29 January 1787, Letter from S. Rudder.
36. *Gentleman's Magazine* vol. 56, p. 926.
37. Ireland, S., *Picturesque Views on the River Thames*, 1792, vol. 1, p. 14.
38. Moreau loc cit.
39. *The Torrington Diaries*, ed. C. Bruyn Andrews, 1934, vol. 1, pp. 259-260.
40. Moreau loc cit.
41. Sapperton Parish Registers.
42. Sapperton and Coates Parish Registers.
43. *Gentleman's Magazine* vol. 56, p. 926.
44. GJ 22 January 1787.
45. Simms, F. W., *Public Works of Great Britain*, division 2, p. 3. Simms & Clark, op cit, pp. 65-6.
46. Storer, J. & H. S., and Brewer, J. N., *Delineations of Gloucestershire* 1824, p. 200.
47. *The Torrington Diaries*, vol. 1, p. 259.
48. See Sutcliffe, op cit, 1816, p. 237; Strickland, op cit, 1826, p. 7; MacDermot, E. T., *History of the Great Western Railway*, 1927, vol. 1, p. 128; Dow, G., *The Third Woodhead Tunnel*, 1954, p. 23.
49. *Gentleman's Magazine* vol. 58, p. 759, and vol. 60, pp. 391-2.
50. Taunton, op cit, p. 265.
51. *Gentleman's Magazine* vol. 60, p. 391.
52. *Gentleman's Magazine* vol. 58, p. 883.
53. Taunton, op cit, pp. 265-9.
54. Dupin, op cit, vol. 1, p. 301. The French edition was published in 1820, the English in 1825.
55. Taunton, op cit, pp. 266-7.
56. For descriptions of the interior of the tunnel now, I am indebted to several adventurous persons, especially C. P. Weaver who made a detailed examination under appalling conditions.
57. Sutcliffe, op cit, p. 222.
58. Devey, J., *The Life of Joseph Locke*, 1862, p. 118.

59. GJ 11 May 1789.
60. *Gloucestershire Notes & Queries* Vol. 4, p. 445, anon. source.
61. *Gentleman's Magazine* Vol. 60, p. 110.
62. GJ 21 March 1785, See also *A Sketch of the Country to which the Abingdon and the Thames & Severn Canals will open a communication with the City of London*, no date (Science Library, South Kensington); and Rudge, T., *General View of the Agriculture of the County of Gloucester*, 1809, plan facing p. 29.
63. Thacker, op cit, Vol. 1, p. 87, quoting from a Cash Book of the Oxford-Burcot Commissioners.
64. JHC Vol. 40, 18 entries between pp. 592 and 926.
65. GJ 27 April 1789.
66. GJ 11 May 1789.
67. The opening ceremony was described in almost identical passages appearing in the *Morning Post & Daily Advertiser*, 24 November, the *Gloucester Journal*, 30 November, and the *Gentleman's Magazine* Vol. 59, p. 1139. An abridged version, evidently from the same source, appeared in *The Times* 26, November 1789.
68. Moreau, S., *A Tour to Cheltenham Spa*, seventh ed. 1793, pp. 158-9, quoting from *The Bath Chronicle*.
69. GJ 4 January 1790.
70. PP 1793 Reports 13/109, p. 31. *Gentleman's Magazine* Vol. 60, p. 110.
71. GJ 1 February 1790.
72. GJ 5 April 1790.
73. *Reports of the Engineers* . . . 1791, Jessop's reports dated 4 and 7 August, 1789.
74. PP 1793 Reports 13/109, pp. 20-1, and see also pp. 15-16, 23, 30.
75. *Reports of the Engineers* . . . , 1791, Mylne's report of 8 May 1791.
76. Mylne, R., *Report to the Gentlemen of the Committee of Subscribers to the Proposed Canal, from Bristol to Cirencester*, 27 September 1793.
77. Priestley, op cit, p. 372. Herefordshire & Gloucestershire Canal, Report of the Committee, 10 November 1796. Minute books in BTR, SHRC 1/1 14 August 1793 & 1 January 1794, HGN 1/1 14 July 1791, Hadfield, C. *Canals of Southern England*, p. 182. Information supplied by C. Hadfield. GJ 9 February 1795.
78. Ireland, op cit, Vol. 1, plates facing pp. 1 & 9, text pp. xii, 1-2, 9, 11-15.
79. Coombe, W., *An History of the River Thames*, 1794, Vol. 1, plates facing pp. 2 & 48, text pp. 30, 44-6.
80. Cooke, W. B., *The Thames*, vol. 1, 1811, and see also first Supplementary vol. of *Views on the Thames,* 1822. The fourth book referred to is Westall & Owen, op cit, 1828.
81. Storer & Brewer, op cit, p. 200.

CHAPTER SEVEN

WHARVES AND WATER

Who designed canal buildings? Architect? Engineer? Local builder? Robert Mylne was a distinguished architect as well as an engineer. Thomas Telford published designs of buildings for canals and turnpikes.[1] Local builders were certainly responsible for some. In the eighteenth century good taste in building was widespread, and, guided by pattern books[2] with designs for mansions, keeper's cottages, coach houses, and details such as roof trusses, windows, doorcases, stairways, local builders designed excellent structures for their patrons, the country gentry,[3] but it was usual for the patron himself, rather than the builder, to engage separately the craftsmen needed — the masons, carpenters, plumbers, bricklayers, plasterers, glaziers — and himself to watch closely the progress of the work. A public company would have found it more convenient to deal with one man who would contract for the whole project, engage the craftsmen, and exercise the supervision. Such a master-builder or building contractor was uncommon before the nineteenth century,[4] and his evolution, like that of the public works contractor, may well have owed a good deal to the rapid spread of the canals.

Certainly the men who built the two largest canal structures in the valley of the Frome, the port offices of the Stroudwater Navigation at Walbridge and of the Thames & Severn at Brimscombe, were of this type, and appear to have designed the buildings unaided by an architect. There seem to have been no pattern books specifically for industrial structures,[5] but there would have been no difficulty in adapting the published designs to suit them.

The Walbridge building was the work of William Franklin. In the autumn of 1795, Franklin attended a meeting of the Stroudwater committee "to receive instructions for a Plan & estimate" for it. When he produced a design for a façade with two wings, the company were very short of money and agreed with him "to compleat the Masonry Work, Plasterers Work, and Tylers Work" of the middle part only. Four months later, however, by which time it was probably obvious that the effect was deplorable, the committee allowed him to add the wings. The building was completed in the spring of 1797.[6]

Brimscombe Port House was the work of Thomas Cook, a native of Painswick, who signed the agreement on 21 December 1786, and incised the date 1789 on the keystone of the office window. A master mason and building contractor who could undertake several large works concurrently, Cook

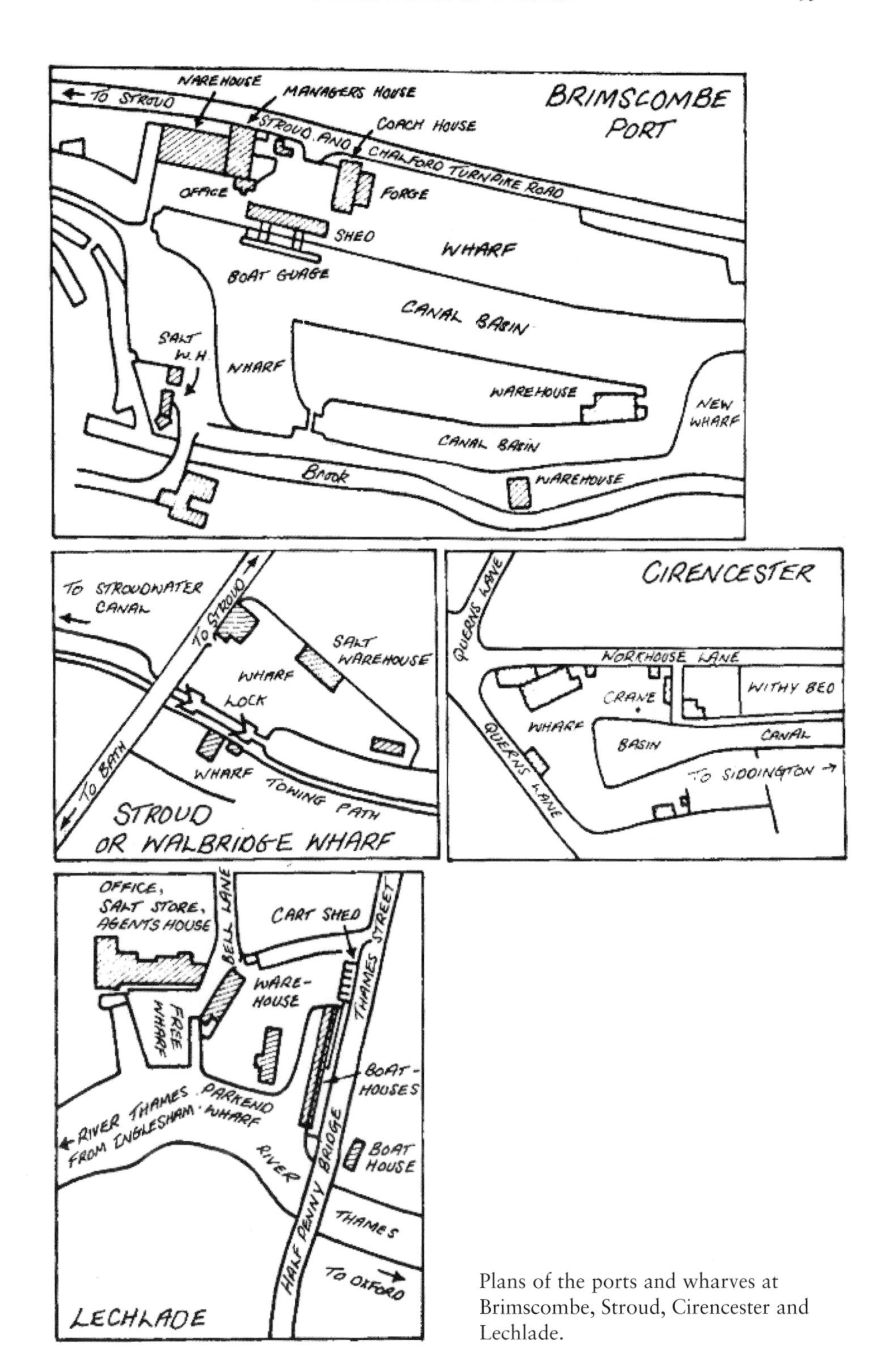

Plans of the ports and wharves at Brimscombe, Stroud, Cirencester and Lechlade.

worked for the Thames & Severn company from 1784 to 1795, was "the contractor" for Gloucester Gaol, built between 1786 and 1791,[7] and obtained the contract for Stafford Gaol early in 1789.[8] An architect designed each gaol, but there is no evidence of one at Brimscombe. The Brimscombe building was by far the largest, and the port installations were the most extensive, on the Thames & Severn. This was because the proprietors did not do as Robert Whitworth had assumed they would, make their canal for Thames barges only. Determined to extend the range of the Severn trows, already carrying coal to Stroud, as far into the manufacturing area as possible, they selected Brimscombe as the highest point where the valley was wide enough to hold the basin, wharves and storage needed for transhipment. It was not in fact the absolute limit of the trows, for Bourn lock, the next above the port, was a hybrid, 90 feet long to take Thames barges and 16 feet wide to suit trows, whereas all further up were only 12¾ feet wide and all further down only 68 feet long.

The port, of which scarcely a trace remains today, fully deserved its name. For many years all cargoes were handled at least once upon its quays; its earnings, swollen by transhipment dues and the trade of a busy area, were more than three times those of all the other wharves combined; it was the focal point of the company's activities and the site of their administrative quarters. The basin, some 700 feet long and 250 feet wide, is said[9] to have been capable of holding a hundred vessels at one time. It was bounded by wharves, and in the middle was an island on which coal was stored so as to be as secure as possible from pilferers. The Great Wharf lined the northern side, and beside it stood warehouse, office and agent's house combined in a single three-storeyed masonry building. A narrow basin adjoined the western wall of the warehouse, enabling a barge to discharge direct into store. A well-proportioned transit shed, dated 1801, stood close to the northern wharf wall and was originally open on all sides with a hipped roof supported on masonry columns. There was a small warehouse on the island; a salt store on the western wharf, its stones so deeply impregnated that to the end salt sweated out on the surface; a forge dated 1794; a sawpit where timbers were cut by two men, the senior standing above to guide the cut, the junior below in the sawdust; and a remarkable feature which few other canals possessed, a boat weighing-machine set up in 1845.

The principal warehouse was severely plain but of great strength. Its floors were carried on joists, beams and stanchions of English oak so tough that modern nails, lacking the strength of the old hand-wrought ones, would never pierce it. The stanchions divided the floors into long aisles 6¼ feet wide and 6¾ feet high, and it was easy to picture the place stacked with great weight and volume of merchandise: sacks of grain, hogsheads of cider, kilderkins of porter, cases of copper, firkins of lead, bacon, timber,

Brimscombe Port from an oil painting made before 1845, showing trows in the basin, approaching from Stroud under sail, and on trow.

Brimscombe Port, the busy scene in 1826.

The massive timbers of the great warehouse in 1947. The floors had been divided up for the use of the Polytechnic.

earthenware, "Punnys of hemp", and carotels of currents are amongst the entries in the bills of lading. The eastern end housed the offices and agent's quarters. From the upper windows of the office the agent could see the whole area of the port, and in the well-lit panelled rooms he and his clerks and apprentices made their flowing and often beautiful entries in the ledgers, registers and letter-books, or pored over the ill-spelt, untidy notes of uneducated employees. A simple device in the screen just inside the outer doorway avoided innumerable enquiries about the depth of water on the summit, often far less than that intended: this consisted of two revolving discs which were set to record the depth on Daneway sill in feet and inches.

The agent's house faced eastwards, overlooking a private garden watered by one of the many springs which rise in the flanks of the Golden Valley. It was designed for a large establishment: the agent and his family, a housekeeper, three or four resident clerks and apprentices, servants, agents from other depots staying overnight on business, members of the committee attending meetings. The house had three storeys served by a handsome "geometry stone" staircase, and contained committee room, assembly room, several reception rooms, twenty-six bedrooms, kitchen, brewhouse, servants' hall, pantry, dairy, two cellars, and, in a separate structure, coach house and stabling for eighteen horses.

Goods were handled at twelve other points along the line: Stroud, Chalford, Daneway, Thames Head, Siddington, Cirencester, South Cerney, Cerney Wick, Latton (for Cricklade), Marston Meysey, Kempsford, and Inglesham. Most of these had at least a 'lay' or widened basin in which a boat could moor clear of the channel, a wharf wall, a warehouse and a stable, but at some there was no more accommodation than the wall of a lock for landing goods and an employee's cottage in which to store them. Daneway saw its busiest days in 1786-89 while it was the head of navigation, but it always had an outward trade in timber and its basin was used by barges waiting to pass through the tunnel. Thames Head wharf, standing beside the Foss Way and expected to serve a wide area including Tetbury seven miles off, was furnished with a pleasant cottage for a resident agent, built and probably designed by the mason John Holland in 1784, a warehouse and four cattle sheds. Siddington was the maintenance depot for the eastern reaches and had a workshop as well as a house. Cirencester, the only terminal basin on the line, Latton and Kempsford, each had a building of similar and odd design dating from 1789: the agent's four-roomed house rose between wings which served as warehouses, and the façade, built of Cotswold rubble stone faced with plaster and stucco, was topped by a triangular pediment and patterned with ashlar quoins.

The port buildings from the island in 1947: in the background, the warehouse, office and manager's house; in front, the transit shed of 1801, later walled between the columns; on the right, the stable.

Chalford wharf and round house photographed by Humphrey Household in 1937. The canal had been closed four years before.

Daneway Basin and cottage, and the road built by the company leading up the hill to Sapperton village. In the bottom right-hand corner is the side-pond formed about 1823.

Kempsford wharf house in 1947, showing the agent's house built in 1789 with wings serving as warehouse space.

The agent at Brimscombe drew a salary of £100 a year and an allowance of £130 to cover housekeeping, cost of linen and glass, keep of horse and cow. Agents were stationed at Siddington (paid £60), Latton (£50), and Kempsford in early days, but the remaining wharves were managed by employees who had other duties as well. There were miscalculations of course: at Kempsford and Thames Head, traffic proved to be much less than expected, and the wharf houses were later let; at Stroud, where the earnings were exceeded only by those of Brimscombe, new quay walls and a warehouse had to be added in 1828; and Inglesham, isolated in the riverside meadows, proved so inconvenient and inadequate as the terminal that in 1813 the company bought Parkend wharf in Lechlade, which had been established by one of the Thames barge-masters about the middle of the eighteenth century. The Thames & Severn increased the accommodation, which included a charming two-storeyed house for their agent, a salt store with an office above it reached by an outside flight of stone steps, a two-storeyed warehouse, a cart-shed of five bays and a dock 190 feet long. It is a scene of activity even now (1969), but the retail coal arrives by land and the boats are pleasure craft.

The warehouses had hoists and the principal wharves cranes, Brimscombe no less than six, one dating from 1793. The early cranes had timber beams, but in 1827 the company's officers studied designs for cast iron cranes to

Latton wharf house seen from the back in 1949, the roof lines showing the peculiar design with the warehouse space surrounding three sides of the agent's house.

Latton wharf, inside the rear portion of the warehouse space.

Inglesham round house, built 1790, warehouse, and the bridge crossing the tail of the lock, in 1923.

Lechlade wharf house 1947, showing the agent's house and the office up the steps above the salt store. Originally built for a mid-eighteenth century Thames bargemaster, it was bought and altered by the company in 1813.

lift five or ten tons, costing £120 or £210 delivered to the site, and the three new cranes set up shortly after at Brimscombe, Stroud and Cirencester were probably of this type, together with another, costing £140, erected at Brimscombe in 1841.

In addition to the staff already mentioned, there were wharfmen at Brimscombe, paid 1s. 6d. a day in 1799 and 11s. a week ten years later, labourers whose pay was 9s. or 10s., and twelve watchmen who were stationed at intervals along the line. These watchmen had to maintain the towing-paths and fences, act as lock-keepers where their lengths included locks, see that boatmen observed the company's bye-laws and report any evasions — but it was well-known that some of them could be silenced by gifts of meat and drink. As the navigation was open from 5 am to 9 pm in summer and 6 am to 6 pm in winter, the watchmen's hours were long even by contemporary standards, yet all but one were paid no more than 9s. a week. The one exception was the Brimscombe watchman, paid 12s. rising later to 18s. a week, but what the company asked of him was humanly impossible — to be on night duty for 9 or 12 hours according to the season (forfeiting 1s. "if found asleep") and to work on the wharf for five or six hours a day. Not until 1813 were his duties divided between three wharfmen, each paid 15s. and each acting as watchman every third night.

The watchmen's cottages were amongst the most interesting buildings along the canal. Some were of a simple rectangular design, such as the gabled example at Sapperton, the particularly charming cottage with a hipped roof at Blue House or Furzen Leaze (near Siddington) which was designed by the engineer James Black and built in 1792, and that at South Cerney which had a bay window giving a good view of approaching boats. But five others, all built in 1790, were circular and a distinctive feature of the Thames & Severn. Even contemporaries were moved by the oddity of these "fanciful round buildings like towers",[10] and after the passage of nearly two centuries they have lost none of their fascination: There is no indication why the design was adopted: certainly it was suitable for its purpose, as excellent views up and down the canal could be had from the windows; certainly it was simple to build, having no quoins. Nor is it clear who thought of it, but there were plenty of local examples of circular buildings which would have been known to any Cotswold mason, some of them dovecots, others towers in which wool was dried, and there was an eighteenth-century example in Lord Bathurst's park.

The five round houses, at Chalford, Coates, Cerney Wick, Marston Meysey and Inglesham, were built of rubble stone covered with plaster and stucco. Of the three storeys, the lowest was used originally as a stable and reached only from outside, the first floor formed the single living-room entered directly through the outer door, the upper was a bedroom. The walls are twenty inches thick, and internally each room is 16 feet 10 inches in diameter, but

The watchman's cottage at Blue House or Furzen Leaze near Siddington, built in 1792, shown as it was in 1950 before alterations.

the curve was slightly flattened on one side to accommodate the kitchen range, the bedroom fireplace, and the stairway which was inserted between inner and outer walls. Two of the round houses carry conical roofs, but the other three are roofed in a curious way: there is a high parapet from which the rafters slope downwards to the centre, where there is a leaden bowl from which a duct leads out through the parapet. At Coates, the entire roof was covered with lead, and the substantial ceiling beams at Inglesham and Marston suggest that these two also were originally roofed with the same material. Possibly the intention was to form a rainwater cistern, particularly valuable at Coates, standing as it does high on the thirsty oolite.

The round houses have their drawbacks: the accommodation is cramped; furniture passes unwillingly through the small doorways and lines the walls uneasily; water is drawn from a well; sanitation is outside. In the closing years of the nineteenth century, an applicant for the post of watchman at Coates visited the round house with his betrothed. She viewed the accommodation with dismay, and when reminded that it had been good enough for others, said roundly that unless it was improved she would not be wed. Faced by such an uncompromising stand, the company gave in, converted the basement from stable to use of the house, built out a kitchen, added another stairway, and partitioned the bedroom.[11]

Disadvantages of the design had in fact been realised much earlier, for in 1831 when it was decided to station watchmen beside the locks at

Coates round house in 1949. At the back can be seen the kitchen built out to appease the bride-to-be.

South Cerney Upper Lock and watchman's cottage in the early 1900s.

Wilmoreway and Eisey, new cottages were built on a rectangular plan with four rooms on two floors above a stable. As the round houses at Cerney Wick and Marston Meysey were then no longer needed, the committee ordered that the first should be let and the second pulled down. Somehow Marston escaped, for in 1835 "a good Tenant offered £5 a year rent" which the committee accepted, but four years later it had become 'very dilapidated' and was again threatened, yet again survived. Indeed, whereas the two cottages at Wilmoreway and Eisey have long been deserted and ruinous, all five round houses were tenanted until recent years, and two, modernised, and cherished for their unconventionality, still are.

In spite of the English climate, it was generally both difficult and expensive to supply canals with enough water. The prevailing method was to divert running streams directly into them,[12] and canal Acts conferred authority to take all waters within a thousand yards or more on either side.[13] But every river and almost every stream turned millwheels which, though running freely from December to June, slowed down and stopped as the springs failed in summer and autumn. Moreover, it had become customary for landowners and tenants regularly to flood water meadows in the early months of the year.[14] Understandably, these established users challenged the companies' widespread powers, and favourable decisions obliged canals to leave enough water for them or else pay substantial compensation.[15]

The more locks, the greater the consumption of water, hence the persistent attempts to invent other ways of surmounting differences in level — boats on cradles ascending inclined planes, boats floating in caissons raised vertically, hoists for cargo containers and railways. Telford and other practical engineers had little faith in any but the last,[16] and indeed several of the devices tried proved to be beyond the mechanical skill of the time except for very small boats, while others had the disadvantage of transhipment. The practical solution was to store water in reservoirs filled in winter, but this was long suspect because knowledge of geology and the theory of dam construction were so deficient that the soil of a chosen site might later prove unsuitable and the dam insecure. Not until William Jessop in 1793 successfully supplied his Grantham Canal entirely from reservoirs[17] was this method widely adopted.[18]

In the meantime, many engineers from Brindley onwards increased the capacity of the summit level.[19] This was what Robert Whitworth did on the Cotswolds, believing that by adding a foot to the depth and six feet to the width he would ensure more water than would ever be needed on the Thames & Severn. What those who adopted this course failed to realise was just how vulnerable the canal then became to leakage from geological conditions or unforeseen accident.

Relying on Whitworth's estimate of the water supplies, the proprietors expected no shortage. Indeed, to those whose knowledge of the area is limited, it appears well watered. In the west, there is the Golden Valley, renowned for its prolific springs. In the east, Thames, Churn, Coln, and countless brooks seam Cotswold with their valleys before uniting among the water meadows. The appearance is, however, very different in a wet autumn than it is after a summer drought, and this also Robert Whitworth failed to realise, for he had made his survey in an October "when the Season was wet".

With the concurrence of Henry, second Earl Bathurst, the Thames & Severn Canal Act had specified that the principal supply of water was to be drawn from the River Churn at the mill-pond of Jenour's Mill, later known as Barton Mill, in Cirencester. The intention was that a feeder about three-quarters of a mile long, partly in the open and partly in culvert, should lead the water to the terminal basin in the town and thence fill the summit level. The Churn, however, turned many millwheels and irrigated Earl Bathurst's meadows; moreover its behaviour is peculiar and its yield disappointing to its users. Fed by the famous Seven Springs and the waters of many streams, the river flows strongly past Colesbourne, but further on, at Rendcomb and North Cerney, it runs over permeable sands and the inferior oolite in which there are 'swilly holes' which yield water in wet weather but visibly swallow it in time of drought. In days gone by, the millers used to repair leaks and consolidate the bed of the river and their mill-ponds by driving teams of bullocks up and down, but even so in a dry season the flow of the river might be reduced by as much as 86 per cent.[20]

As soon as the company opened negotiations for the purchase of Jenour's Mill and its water rights in October 1785, Earl Bathurst went back on his agreement, and a prolonged controversy began. Torn between the needs of himself and his tenants on the one hand, and the statutory rights of the canal on the other, he vacillated from compliance one day to repudiation the next, and finally threatened that if he could not use the water to flood his meadows, "he would do everything in his power to Stop the Works". Nevertheless, the company completed the feeder and drew water through it, whereupon the dispute developed into a ludicrous struggle for the millpond. Earl Bathurst sent men to build a stank across the intake of the feeder. His son Lord Apsley, his agent and two gardeners threw down the banks and did "other mischief": Samuel Smith read aloud to them the section of the Act threatening with transportation any person guilty of damaging the canal works. Perry and Chambers ordered Bathurst's men to desist. But the stank was made and the water cut off.

After six months without this supply, the company diverted the Daglingworth Brook, paying John Pickston 2s. a night for six nights to guard against further interference, and setting up a spoon, worked by John Nock's horses, which ladled water into the feeder for four weeks. Then at last in May 1787 Earl Bathurst conceded the company's right to the Churn water, leaving the question of compensation to be settled later — much later as it turned out. Three attempts to negotiate terms proved abortive, so the Churn millers, who were by then as eager as the company to reach a permanent settlement, convoked the commissioners. But although twice summoned, the latter never met, and their failure to do so arouses the suspicion that they were not prepared "without Favour or Affection, Hatred or Malice, truly and impartially"[21] to ensure that the provisions of the Act were carried out and make an award distasteful to Lord Bathurst.

The commissioners having failed them, the parties resorted to arbitration by two independent millers, Samuel Twamley of Bromsgrove chosen by the company, and John Hancock of Stratton by the Churn millers. In May 1791, the two men settled the rates of compensation and the periods during which these should be paid to eight mills between Cirencester and Latton, the lowest one six miles below the intake to the feeder. The company were not required to pay anything when there was plenty of water for all users, nor when water drawn at night for the canal left enough to work, the mills the following day; but when water was short and the company took the whole flow of the river, they had to pay a specified amount per hour based on the quantity of grain each mill was capable of grinding in that time under normal conditions. Shortly afterwards, the company bought Watermoor Mill in Cirencester, and that affected yet another user, Cripps' New Mills, so a subsidiary agreement was concluded on the same lines as before, and thereafter the total rate of compensation payable under the award was 6s. 5½d. an hour. The maximum cost to the company in any one year was £535 11s. 2d. in 1819, but the average cost was about £350 a year. The terms were always regarded as being particularly reasonable, and nearly a century later, by which time water-milling had decayed and Barton Mill itself was driven by steam, it was considered unwise to seek a new agreement lest the millers' compensation should be increased. As mills went out of business, the rate of compensation of course decreased, but as late as 1904 there were still five receiving 3s. 6½d. an hour. The Thames & Severn was more fortunate in this matter than many others, for, as the committee of the Herefordshire & Gloucestershire Canal remarked a few years later, "the unreasonable demands, always made by Millers on Canal Companies, are too well known to need any comment".[22]

The western end of the summit and the locks in the upper part of the Golden Valley were expected to be filled from the River Frome and springs in Sapperton parish, but these, like the Churn, proved to be of only limited value. For the sake of the millers, the company's use of the river water was restricted to Sundays, and even as late as the end of the nineteenth century when all the mills above Chalford were either deserted or decayed, it was only on sufference that the canal was allowed to draw water on weekdays. As for the Sapperton springs, Earl Bathurst succeeded in making the company's use of these dependent on his pleasure. A reservoir in the upper part of the valley was therefore necessary from the first. One was made at Daneway, but the ground proved to be unsuitable and the supply limited, so it was abandoned and another established at Baker's Mill. This was six feet deep, nearly 300 yards long, and held more than 3¼ million gallons, and the company assured its supply by buying Puck Mill above it. At Chalford, a mile-and-a-half below, a spectacular cluster of some twenty prolific springs known as The Black Gutter, supplied the remainder of the canal with such abundance that the water gives deserted reaches an air of expectancy even now.

Eastwards of the summit, other sources were tapped — and some of the gravest mistakes made. Two miles below Siddington, the Boxwell Springs break out in a withy bed, and when the canal was built it was supposed that these would fill the pound above the two locks at Wilmoreway, which had a combined fall of 18½ feet, altogether inconsistent with those on either side. The reach was therefore widened to form a reservoir, but someone had miscalculated badly, probably through gauging the flow in the rainy season, for the main outflow of the springs was twelve feet below the level of the canal and not powerful enough to rise that height even if contained within puddle banking. As early as 1792 the pound was reduced in width and its level lowered 3½ feet by the insertion of Boxwell Spring Lock, built by Thomas Cook at a cost of £86 14s. 6d. As Taunton put it,[23] "some less copious springs languidly flow" into the canal below the lock, but their supply was wholly insufficient to sustain the operation of the two locks at Wilmoreway which, until modified many years later, were liable to drain the canal at the passage of every boat.

Further east, the company incurred the displeasure of another landowner who wanted to flood his water meadows. The pound between Latton and Eisey locks had been cut through gravel and leaked, but Clowes believed that all would be well in a year or two, so the company decided to take a temporary supply of water from Down Ampney Brook. This was on Lord Eliot's land, and very unwisely they began to cut a feeder without asking his leave. As they might have guessed, he stopped the work immediately, sent his agent to destroy the dam at the intake to the feeder, and once again summoned Smeaton for advice. Now Smeaton had discussed the supply at this very point with Whitworth before the passing of the Act and his advice had not been followed, so neither he nor Eliot minced their words. Smeaton criticised the unequal lock falls from Siddington downwards, and Lord Eliot, in forwarding to the canal proprietors a copy of Smeaton's report, commented with relish upon "the Want of knowledge in the very first principles of Canal making" shown by those who had built the line. The company stood firm, however. They notified his lordship that they intended to exercise their right to the water, "leaving themselves open to the damages they may necessarily commit", and conveyed a delicate warning against any repetition of his agent's interference with their works. Time proved that the need of the feeder was far from temporary, so an agreement was concluded with Lord Eliot in 1796 allowing the company to take the supply from 1 June to "the first Rains in the fall of the year", which did not deprive his lordship's tenants of their precious floodwaters.

The last supply, and a very valuable one which fed the Five Mile Pound between Eisey and Dudgrove locks, was the Whelford Feeder, about 1¾ miles long, drawing from the Coln two miles below Fairford. The course of this lay partly through fields, and when the committee gave orders for the work to be begun in May 1789, they directed with characteristic consideration that where the line lay across land already seeded with corn, no more should be done than to mark it out, and in fact the feeder was not completed until after the harvest.

A single summer was enough to convince the company that their water supplies were wholly inadequate and that, like many another,[24] they must find an entirely new source of proved reliability. They had not far to look: an abundant one which they had not tapped lay midway along the summit in the meadows of Thames Head, where rose "the numerous little fountains" and "various small jets d'eau" beloved of contemporary topographers and artists,[25] amongst them Joseph Farington R.A., who painted the Thames Head Well in Trewsbury Mead, west of the Foss Way,[26] a spot considered by many to be the true source although in fact neither then nor now does the visible source have a fixed site. Normally water appears west or east of the Foss Way within a length of about half a mile, but in extreme seasonal conditions it may rise as much as 600 yards further up, or more than a mile lower down, depending on the level of the water-table in the permeable great oolite rock. The last point at which water would fail in the valley is where the great oolite dips below the forest marble, and there stands Lyd Well, "held by some to be Roman".[27]

Thames Head Well from a print dated 1793 engraved from a painting by Joseph Farington, R. A. Above the wall bordering the canal towing-path can be seen the mast of a passing barge.

Some of these facts, if not their explanation, would have been known to James Black when he and other engineers closely examined the Thames Head springs in the middle of 1790 — he might indeed have been one of the very men depicted by Joseph Farington making notes beside the well. Two years earlier, the company had begun to explore the area to the east of the Foss Way by sinking pits on both sides of the canal. One of these pits had been sunk at a point then known as The Firs, somewhere near Lyd Well, and too far away to be of any immediate use, but another was close to the canal on the north side. This had been sunk to a depth of 54½ feet, whence a borehole had been driven some fifty feet deeper through the permeable great oolite and the impervious fuller's earth to dry rock, and here the company had installed a "Wind Engine", or six-sailed windmill, to work a pump of eleven-inch bore. It was sketched in the summer of 1790 by Samuel Ireland, who recorded that it was capable of throwing up "several tons of water every minute".[28] So it was evident that a reliable supply could be obtained by sinking a new well deep into the great oolite and installing a powerful pump to lift the water into the summit level,[29] but before more could be done, the consent of the Thames Commissioners had to be obtained and the best possible site found for the well. The commissioners gave their consent on condition that the canal would return the water to the river at Inglesham.

The first pump at Thames Head was a six-sailed windmill, replaced in 1792 by this Boulton & Watt single acting beam engine, working a much deeper well, which remained in use until 1854. It is depicted as from the road in this view of 1828.

The plan for a dependable installation was then evolved by James Black in collaboration with James Watt, who himself visited Thames Head, and on 1 January 1791, an indenture was signed authorising the company to set up a 'fire engine' or steam pump on the south side of the canal a little to the east of the Foss Way, on condition that they paid Boulton & Watt a premium of £120 a year for the patent rights. It was a bold move, for although several canals had used Newcomen engines, only the Birmingham Canal at that date had installed Watt's;[30] moreover, whereas the Birmingham had plenty of coal close at hand, all that used at Thames Head would have to be brought from far off.

Watt's single acting beam engine designed for the Thames & Severn used steam at a pressure of 3 lb per square inch in a vertical cylinder 50 inches in diameter. The piston had a stroke of 8 feet and worked a horizontal beam, pivoted at its centre, from the other end of which hung the pump rodding. Steam above and a vacuum below drove the piston down and operated the pump. Steam then circulated freely on both sides of the piston, so that the weight of the pump rodding performed the return stroke, actuating as it did so the mechanism which drew the steam from below the piston to the condenser and set the engine ready for the next stroke. The engine was of 53.2 hp and designed to work a pump 26 inches in diameter lifting water 70 feet at the rate of ten strokes a minute.[31]

As Boulton & Watt did not at that date themselves manufacture complete engines, it was usual for them to supply a set of drawings (there were 39 in this case),[32] some of the smaller parts, and the services of an experienced erector, and to order the principal castings (cylinder, condenser, and pump barrel) from either the Carron Ironworks, John Wilkinson, or Reynolds of Coalbrookdale.[33] For the Thames Head engine, most of the larger parts, including the boiler, were supplied by Wilkinson, but some material was obtained from Reynolds. The erector was Isaac Perrins, a well-known prize fighter, and under his direction were the company's carpenter Benjamin Haines, who was paid 15s. a week, a smith Thomas Toward, and up to a dozen of Thomas Cook's men who built the stack and no doubt also the engine-house. Toward came from the north and was probably already an experienced engine-builder as he was paid 21s. a week and afterwards stayed on as engineman at a salary of fifty guineas a year, which was well above the contemporary average.[34] Toward served long and faithfully, tending the engine and doing repairs, as he put it "boath early and whol nights & Sundays when necessity required it". His name and the date 1792 were incised on one of the stones of the engine-house and survived among the ruins until a few years ago.

There seems to have been little realisation of the time it would take to set up the pumping station. The Boulton & Watt indenture mentioned erection of the engine in eight months, and the agreement for sinking the well, made with a miner named James Philps, specified six weeks. In fact, it all took 3½ years.

Stone inscribed with the name of the first engineman, T. Toward 1792, among the ruins of the pumping station in 1948.

The beam, 24 feet long, 43 inches deep and 39 inches broad, was made from a Kelvedon oak which cost £91 8s. 6d. delivered to Limehouse Wharf. It was set up in October 1791, and was an occasion for celebration. Perrins asked for beer for himself and Toward, and they were given a guinea between them, but Haines, whose responsibility for the timber beam was certainly greater but whose tastes were evidently different, had a guinea to himself.

The engine was started on 22 September 1792, probably to help the well-sinkers. The great oval well, 15 feet by 10 feet clear and lined with a masonry curb 2 feet thick, was sunk by Philps for 21s. a yard in depth, the company providing him with windlass and rope but he himself finding tools, candles and powder. When completed early in January 1793, it was 63 feet 8 inches deep. The springs entered 47 feet below the top, and the supply was increased by galleries driven out from the sides. The most important of these galleries, known as The Little Tunnel, left the well at a depth of 55 feet 8 inches and communicated with two other wells, beyond which it was to be carried under the floor of the valley. But even in the dry season it was not always possible for the miners to work there, and in 4½ years Philps only drove The Little Tunnel some 116 yards, after which twenty-two years passed before there was an opportunity to resume work.

The rising main extended nearly to the bottom of the well. Forty feet down was the pump barrel, and below that the valve or 'clack door', opening upwards. The bucket, or piston of the pump, was fitted with two flaps of thick leather, reinforced with iron plates, which opened upwards as the bucket descended. Water was therefore raised by suction through the clack door into the pump barrel, and thence by direct lift of the bucket, successive strokes building up a column of water to the top of the rising main. The nominal stroke of engine and pump was 8 feet, but unlike a rotative engine whose stroke is governed by the throw of its crank, the working stroke of a beam engine depended on the engineman, who varied the setting of the valve gear to suit the depth of water in the well. If that level was very low, he took control of the valves himself (as Toward many times did), working them by hand and watching closely for any sign of the pump drawing air instead of water, and ready, should it do so, to stop the engine before the combined weight of the rodding and the column of water in the rising main should drive the unsupported bucket downwards with a violence that could severely damage the entire installation.

The pumping station cost £4,015 to set up, and was brought into service on 14 June 1794. The normal working season of the engine, pumping 12 to 24 hours a day as the supply from other sources diminished, was from June to the breaking of the springs towards the end of the year. It supplied about one lock-full of water an hour, and its performance, or 'duty' as then measured in terms of pounds raised one foot per bushel of coal, was 15,716,000, about

Thames Head pumping station showing the Cornish engine acquired from Wheal Tremar near Liskeard, which replaced the earlier and by then worn-out pump in 1854.

average for Watt engines in Cornwall at that date. It had been built, according to John Taunton, "at a late period of Watt's experience in the improvement of the Steam Engine", when design had reached a state almost as perfect as possible with the use of low pressure steam. Within twenty years, however, the far more efficient high pressure pumping engine was introduced in Cornwall, but although the Thames & Severn company chafed at the quantity of coal consumed at Thames Head — pumping cost about £480 a year, exclusive of interest on capital and the payment of Watt's premium — it was not an economic proposition to replace the old engine, especially as it was capable of pumping more water than was then available in the well. Indeed, in spite of setting up the pumping station and spending some £830 a year to draw water from the well and the River Churn, the canal did not have enough to meet demand until further works were completed more than forty years later.

As the years passed, the pumping station was blamed for the failing springs and dry watercourses at Thames Head.[35] Yet one observer, watching the working of the engine in a season of drought, recorded that the depth of water, in the well was not reduced at all, and Taunton found by long experience that even continuous pumping in similar conditions only lowered the level by three or four inches in a week.[36] The critics were perhaps unaware of the statement John Leland had made 250 years before the canal was built, that at "the very Hed of Isis in a great Somer Drought apperith very little or no Water",[37] the truth of which is just as apparent now, many years after the canal has been abandoned.

REFERENCES

1. Telford, T., *Life*, Atlas of Plates.
2. Halfpenny, W., *Useful Architecture in Twenty-one New Designs*, 1752; Morris, R., *Select Architecture*, 1757; Pain, W., *The Practical Builder*, 1774.
3. See Lloyd, N., *A History of the English House*, 1949, pp. 115, 153; and Richardson, A. E., *An Introduction to Georgian Architecture*, 1949, pp. 28-9.
4. Postgate, R. W., *The Builders' History*, 1923, pp. 9, 72-3; Cf Clapham, *The Early Railway Age*, 1926, pp. 162-4, 595.
5. The Librarians of the Royal Institute of British Architects and of the Royal Society of Arts are both unable to recall having seen anything of the kind.
6. SWN committee vol. 2, pp. 17, 202-6, 213, 240.
7. Proceedings of the Commissioners for building Gloster Gaol (GCRO). GJ 20 February 1786.
8. Roxburgh, A. L. P., *Know your town — Stafford*, 1948, p. 132.

9. Storer & Brewer, op cit, p. 202.
10. Rees, *Cyclopaedia*, Canal.
11. The story was told by Delderfield, E. R., *Travel with me*, 1949, p. 133. The alterations could be seen.
12. See, for example, Frisi, P., *A Treatise on Rivers and Torrents . . . . To which is added, An Essay on Navigable Canals*, 1762 (English ed. 1818), p. 168; and Vallancey, op cit, p. 131. Both these writers envisaged running streams as the principal supply.
13. Two thousand yards in the case of the Thames & Severn Canal (Act 23 Geo. III cap 38 section 1). Rees, Priestley and Phillips mention distances up to two miles.
14. See the contemporary reports entitled *General View of the Agriculture of the County of . . .*, particularly Gloucestershire, Hants, Sussex.
15. Hassell, J., *Tour of the Grand Junction*, 1819, pp. 37-8.
16. See the comments of Phillips, Sutcliffe, Telford and Chapman (Observations on the various systems of Canal Navigation, 1797). It is significant that Sutcliffe mentions only railways as an alternative to locks.
17. See Jessop's Observations on the use of reservoirs for flood waters written in May 1792 and published in Pitt, W., *General View of the Agriculture of the County of Stafford*, 1794, pp. 39-44. See also Hughes, S., *Memoir of William Jessop*, pp. 22, 32; and Sutcliffe, op cit, 1816, pp. 98, 109-110, 176; Hadfield, C., & Skempton, A. W., *William Jessop, Engineer*, 1979. pp. 56, 260.
18. The pages of Phillips and Priestley mention reservoirs in connection with some 14 navigations up to the end of 1792, and some 32 subsequently, no less than 12 of which were existing canals seeking to augment their water supply.
19. Priestley, op cit, pp. 187, 210, 387, 425. Rees, *Cyclopaedia*, Canal. Sutcliffe, op, cit, pp. 98, 176.
20. Richardson, L., *Wells and Springs of Gloucestershire*, 1930, pp. 42-3. Taunton, J., *Some Notes on the Hydrology of the Cotteswolds* in *Proceedings of the Cotteswold Naturalists' Field Club*, vol. 9, p. 59.
21. Minute Book of the Commissioners appointed under the Act of 1783 (GCRO).
22. Herefordshire & Gloucestershire Canal Report of the Committee, 10 November 1796 (GCL).
23. Taunton, J., *Boxwell Springs* in *Proceedings of the Cotteswold Naturalists' Field Club* vol. 9, pp. 59-60, 70-71.
24. See Sutcliffe, op cit.
25. Fearnside, W. G., *Tombleson's Views on the Thames and Medway* c. 1834, pp. 7-8.
26. Coombe, op cit, vol. 1, p. 3, and plate facing p. 2.

27. Richardson, op cit. pp. 36-7, 106.
28. Ireland, op cit, vol. 1, p. 2, and plate facing p. 1.
29. Robert Mylne in his report To the Gentlemen of the Committee of Subscribers to the Proposed Canal from Bristol to Cirencester, 27 September 1793 (collection of C. Hadfield) claimed that the summit of the T & S canal could have been 50 feet below the level chosen and pumping at Thames Head thereby eliminated. But he was at pains to refute opinions expressed by Robert Whitworth and Clowes and to criticise their work. Some of his criticisms were valid but would he himself have faced the prospect of making Sapperton tunnel at least 1½ miles longer in the east where he would have had to drive it below ground water logged in winter?
30. Boulton & Watt Collection (BRL), Engine Book Section C. 54 canal engines are listed, five of which had been ordered by the Birmingham Canal between 1777 and 1790.
31. Boulton & Watt Collection (BRL) and T & S Collection (GCRO).
32. Boulton & Watt Collection.
33. See Dickinson, H. W., and Jenkins, R., *James Watt and the Steam Engine*, 1927, pp. 256-7, 263, 270, 346. See also Raistrick, A., *Dynasty of Iron founders*, 1953, pp. 93, 149-150, 152-3.
34. According to Ashton (*Iron and Steel in the Industrial Revolution*, 1924, p. 192), the average wage of enginemen in 1786 was 15s. or 16s. a week.
35. See Fearnside, op cit, c. 1834, p. 10; Hall, Mr & Mrs S. C., *The Book of the Thames*, 1859, p. 14; Beecham, K. J., *History of Cirencester*, 1886, p. 283. See also Richardson, op cit, p. 106, where various opinions are quoted.
36. Richardson, op cit; p. 106, quoting John Bravender and John Taunton.
37. Hearne, T., *The Itinerary of John Leland the Antiquary,* 1769, vol. 5, p. 64.

CHAPTER EIGHT

"CONDUCTING A COMPLICATED CONCERN"

There is little real knowledge of how canal carrying was first organised, but it has been generally supposed[1] that the canal companies themselves did not do it: Halsbury's *Laws of England* stated that "most Canal companies originally had no powers to engage in the business of carrying",[2] and the Act of 1845[3] expressly giving them those powers seemed to reinforce that belief. Yet the Thames & Severn Act, like others, stated the right of "all Persons whomsoever" to use the canal on payment of toll,[4] and contained no suggestion that the company alone were to be denied that right.

Clearly someone had to provide the capital to build and equip sufficient numbers of vessels to operate on any new canal, and also to finance development of the coal trade in areas not previously touched. It is unlikely that the carriers already engaged on the neighbouring river navigations were in a position to do so: as has been shown, those on the rivers Thames and Severn were in a very small way of business, and in fact very few of them took to carrying on the Thames & Severn in its early days or, when they did so, found much profit in it. It seems unlikely that conditions on other rivers and canals were very different.

On the other hand, there is plenty of evidence that canal proprietors provided capital for both carrying and coal trading. Some did so individually, as general carriers or for the carriage of their own special products, for example, the Duke of Bridgwater on his own canal, Josiah Wedgwood on the Trent & Mersey,[5] Benjamin Grazebrook on the Stroudwater and River Severn.[6] Others formed a subsidiary company, such as Hugh Henshall & Co. whose business was run from the office of the Trent & Mersey at Stone.[7] In quite a number of cases, however, the proprietors engaged in carrying quite openly in their corporate capacity, notably on the Forth & Clyde,[8] Basingstoke,[9] Herefordshire & Gloucestershire,[10] and Thames & Severn. The Birmingham,[11] Stroudwater[12] and Stourbridge[13] companies, all traded openly in coal.

Yet by 1825 all canals, it is said, had "found the inexpediency of continuing to be" carriers, and most had given it up.[14] There can be little doubt that this was due partly to the complications introduced into the administration in an age with little experience in the management of large and dispersed businesses, and partly to loose accountancy, such as failure to charge full tolls to the company's own boats or to debit their trading account with the cost of the vessels, which together could long conceal whether the venture was profitable or not. The proprietors of both the Thames & Severn and the Basingstoke[15] companies, for example, discovered with a shock after many years that their carrying was un-remunerative.

After so much activity, how then did there arise the belief that canal companies had no power to act as carriers? The most likely explanation is that there was an action at law — or at least the threat of such action challenging the operations of some body of proprietors on the grounds that they had acted unfairly by failing to debit the accounts with full tolls or warehouse charges, thereby tending to establish a monopoly.[16] Any such proceedings would have acted as a deterrent until the general enabling Act was passed in 1845, by which date all fear of monopoly had been dispersed by railway competition.

The Thames & Severn company's first need of boats was to deliver constructional materials, and for this they bought second-hand Severn craft, such as the frigate *Nancy* for £24 in 1784. Very soon after they ordered from William Bird, a Stourport shipwright, a 60-ton trow, *Littleton*, and a 25-ton frigate, costing £135 and 50 guineas respectively, and began building their own boats, engaging William Large, "carpenter and millwright", to do so. Large built two 30-ton canal boats, *Endeavour* and *Adventure*, on a temporary site at Walbridge, and then in 1786 completed two 60-ton Thames barges, *Severn* and *Thames*, at The Bourn above Brimscombe.

This was the site chosen by the company for a permanent boat building yard, and under their enterprising direction it became the scene of exceptionally interesting development of the traditional river craft of both Severn and Thames. Moreover, under the company's successors barges were built there until well into the twentieth century. It was James Perry who blazed the trail by placing an experienced boat builder in charge, Samuel Bird, probably William's son. Bird's first vessel was a Thames barge, appropriately named *Experiment*, completed in 1790 and the prototype of many others built to his specification, which listed all the timber and ironwork required and estimated the cost, including labour, as £160 16s. 10d. In size, *Experiment* probably differed little from the company's previous Thames barges, which were 80 feet long, 12 feet wide, with a capacity of 70 tons and a laden draught of 4 feet, but whereas the traditional western barges in use on the Thames had straight sides and punt-shaped ends, Bird's new ones were of greatly improved shape. Fortunately an engraving published in 1793 shows one approaching Sapperton, and the bluff rounded bows, curved sides and square stern show the strong influence of the Severn craft with which Bird had been familiar. Other barge builders probably adopted the design quite soon, for in 1795 five new Thames barges were ordered from riverside yards at Abingdon, Pangbourne and Reading, and it was presently reported that they would be "rather more contracted at bottom than those already built". These took four or five months to complete.

His next vessel was a salt-barge built in 1791 for the Droitwich trade of a private owner. Salt-barges, 60 feet long and 12 feet wide,[17] were already in use, and such was the importance of the trade that they were the only vessels of less than 60 tons burden which were allowed to pass through the locks in the season of short water, and the only ones trading direct from Severn to Thames. After measuring the length of every lock between the Severn and Brimscombe and experimenting in the tunnel with a barge carrying a temporary frame, the *Droitwich* was built to the maximum dimensions possible: 65 feet long, 12 feet 2 inches wide, with a burden of 40 tons of salt or 50 tons of "any Solid weight" on a draught of 4 feet. She cost £150.

There followed the most ambitious of all the company's vessels, one intended to trade between west Wales and Brimscombe without transhipment. Her design posed a problem because the open waters of the Bristol Channel were beyond the range of a square-rigged flat-bottomed trow, yet no vessel with an ordinary keel was safe in the river or possible on the canals. James Black, however, knew of the recent invention of a 'sliding keel' of centre-board type, made of oak planks and raised and lowered by a winch, which had been fitted in a flat-bottomed cutter of 120 tons built at Plymouth in 1790. Drawing only 6½ feet with the keel raised instead of the 13 feet normal for her size, this

Engraving dated 1793 showing a Thames barge of the company's improved design approaching the tunnel at Daneway.

cutter appeared to justify the inventor's claim that she would be able to sail direct from ports on the wider canals "to the most distant parts of the known world".[18] For the trade in copper from Swansea and malting coal from Tenby, Black therefore designed the *Swansea* of 62 tons, 61 feet long and 15 feet 1 inch wide, which was completed at The Bourn in April 1793 with a sliding keel and schooner rig.

The carrying and trading had been begun in 1788 by a select committee of proprietors, including amongst others, Littleton, Loveden, Perry, Chambers, John Chalié and James Black, who after six months formed James Perry & Co. with Perry himself giving characteristically energetic lead. In course of time, however, there arose some doubt who benefited most from the peculiar financial arrangements between the two companies, and eventually management of the whole carrying and trading business was put in the hands of a committee of trade.

The fact is that while there was trade between the canal and the Severn, there was very little between the canal and the Thames. Only 112 boats passed over the summit in the second half of 1792, and even with the help of concessions the carrying did not pay. Encouraged by a reduction of toll on goods carried the full length of the canal, a group of nine individual barge-masters was persuaded to experiment with a fortnightly service to London, carrying any goods except household furniture for 23s. 6d. a ton from Brimscombe to London and 28s. a ton on return, exclusive of canal toll. But the arrangement did not last long, and when the company tried it for themselves in 1793 they soon found that expenses exceeded receipts.

Nevertheless, it was no good having a canal without using it, so the committee of trade, backed by the proprietors, applied to the expansion of the business all the resources they could lay their hands on, including a great many private loans advanced by individual members. They ordered more boats, appointed local sub-committees, arranged for barges to berth in London at the Hambro Wharf of Jonathan Sills & Sons, merchants and wharfingers, appointed Joseph Sills as superintendent of trade at an office in Upper Thames Street, installed a navigation agent at Brimscombe, and opened agencies in Bristol, Gloucester, Tewkesbury, Upton-on-Severn, Worcester and Stourport. Some of these agents were salaried servants, others were remunerated by the wharfage charges and a commission of 4d. in the pound on freights, an arrangement wide open to abuse as the Worcester agent proved by pocketing more than 70 per cent of the company's share as well as his own; having "acted in a manner much unlike an honest man" (so the committee noted), he landed in prison.

Under the impetus of this expansion, the company's staff and expenditure grew rapidly. A Thames barge, for instance, carried a crew of three, each paid £5 or £7 a trip in addition to their keep. The barges generally sailed in pairs in charge of a cost bearer who paid all expenses, and so found it easy to inflate his accounts and line his pockets.[19] At least one was dismissed, as also was a clerk stationed at Abingdon to check the cost bearers' accounts. Then there was labour for towing. Below Brimscombe, men had to do the whole of it, and even above they did most of it, towing, as shown in an engraving of 1793, by a line attached to the top of the mast, earning from 10d. to 1s. 6d. per ton. One can readily appreciate that whenever the wind was favourable a square sail or a sprit-sail was hoisted to ease the manual labour. Only one stud of towing horses appears to have been maintained by the company, and that was for use on the upper Thames between Lechlade and Abingdon.

Through the tunnel there was no towing-path, and as the use of "Shafts, or Sticks, or other Things, against the Arch of the Tunnel" was prohibited, men had to leg it against the roof and walls, lying on their backs on top of the cargo and on the narrow deck at the side of a Thames barge (which closely fitted the bore), or on detachable wings projecting from narrower craft. There were occupational hazards: men were known to fall off and meet a dark and terrifying death,[20] and in course of time they often developed a lumbar complaint known inelegantly but expressively as 'lighterman's bottom'. At best, progress was painfully slow, taking some five hours eastbound against the flow from the pumping station, and three hours westbound. Moreover as barges could not pass in the tunnel, entry at either end was restricted to fixed times and no more than three passages in each direction were possible in twenty-four hours. Long delays therefore occurred before the tedious legging even started, delays which had been even greater before the company yielded to pressure and allowed the tunnel to be used by night as well as day, for they, like most others, closed the canal entirely at night.[21]

With their growing labour force, the company faced a difficulty common to all employers in the latter part of the eighteenth century: that of finding small coins with which to pay wages. For years not nearly a sufficient number had been struck by the Mint, and the alternative to paying groups of employees in notes which they afterwards cashed, probably at a discount, with shopkeepers or publicans, was for employers themselves to issue tokens for local use. Some tokens were almost worthless, but the halfpennies struck for the Thames & Severn in 1795 were of good quality, and many surviving examples are well worn with use that proves local confidence. The obverse shows a boat under sail, probably one of the company's Severn trows, the reverse the eastern portal of Sapperton tunnel, and around the edge is the inscription PAYABLE AT BRIMSCOMBE PORT. There are three varieties of copper token and one of silver, between which there are only minor differences, but a fourth struck in copper showed striped sails on the trow instead of plain ones and is "very rare, as the obverse die failed almost immediately".[22]

From 1798 Thames & Severn barges plied regularly once a week between Brimscombe and London, but travelling only by day, they took nine or ten

Obverse of the copper halfpenny tocken struck for the company in 1795.

Reverse of the token showing the Coates portal of the Sapperton tunnel.

days to reach London and from twelve to fourteen to return, even in the most favourable conditions. There were also "Thames and Severn Canal Market-Boats, which leave Cirencester every quarter-day of the moon", the Bristol trows which connected with these, and the Stourport service which gave delivery of goods from London in fourteen to twenty days. Arrangements with other carriers were advertised so that goods could be forwarded to a wide range of places in various directions, such as down the Bristol Channel to Bridgwater and Newport, westwards to Hereford, up the Severn to Shrewsbury, and through other canals to Birmingham, Manchester and Liverpool.

Passengers? Other than the scenery, there was little to attract passengers to a route tediously slow via circuitous rivers, so many locks and a long tunnel any stage wagon could do better. There is no reference to regular passenger services such as operated on some canals; in fact, the only reference at all suggests occasional market folk or hitch-hikers in a bye-law forbidding barge-masters to take aboard "any Person . . . to be Passenger . . . without Leave in writing from some of the Wharfingers and paying for the same".

Mileage was the basis of all charges, and almost all canals (the Stroudwater was an exception) had therefore to set up milestones along the line. The Thames & Severn used heavy rectangular cast iron plates secured to large stones and recording the miles and half-miles from Walbridge and the alternate quarters from Inglesham. The charges were twofold, a sender paying freight to the carrier and toll to the canal company. Tolls varied greatly between one canal and another, but the general principle was that each commodity should pay according to its value 'charging what the traffic will bear' as the railways did for so long. Some cargoes were allowed to pass toll free: government stores perhaps, or roadstone, or on the Thames & Severn, manure for fields if it did not pass through a lock. Thames & Severn tolls were 2d. or 3d. per ton per mile, except in the case of coal, which was charged in zones (the lowest about 1¼d. per ton per mile) with complex arrangements in the Stroud area interlocking with the Stroudwater company's high toll and intended to prevent competition; the effect of these was that coal landed at wharves below Brimscombe paid 3s. 6d. a ton, the same as on the Walbridge wharf of the Stroudwater, whereas that landed at Brimscombe paid only 2s. 3d.. As one would expect, there were frequent complaints from Walbridge that coal bought at Brimscombe Port was reaching hearths and furnaces in Stroud.

While the Thames & Severn acted as carriers, they quoted a number of rates including both toll and freight: for example, nails from Stourport to London 34s. a ton, bacon from Bristol to London 37s. 6d. a ton, and cheese from Gloucester to London 38s. a ton — less than half the cost of carriage by land.

The London wharf accounts show a great variety of consignments loaded and unloaded as the company's boats lay at Hambro Wharf: oil in runlets, hogsheads,

Milestone.

The effect of straining towing-lines on a curve, scouring to a depth of three inches in this stone protecting the abuntment of Thames Head Bridge, as it was in 1949.

Sketches of merchandise by W. H. Payne, 1804.

puncheons and pipes; puncheons of the perry for which Gloucestershire was famous; sugar in lumps, loaves, bags and hogsheads; firkins of butter; bobbins and 'matts' of flax; pockets and bags of hops; casks of purgative squills; sticks of timber; a tierce of alum; "8 Serons of Barilla", crates or hampers of a vegetable alkali from Spain which was used in the manufacture of soda, soap and glass. But the principal commodities, according to evidence given by James Black in 1793,[23] were tin, iron from Coalbrookdale, cider, copper, lead, salt, cheese, brass, aquafortis from Bewdley, all of which were carried to London, and from London, all kinds of groceries, rags, oxhides, tallow, dye-woods, potash and oil, the last three for use by the clothiers. There was also a considerable trade in grain from Lechlade, Cricklade and Cirencester to Bristol, and of course coal from the Severn.

Under the stimulus of the company's services traffic grew: 420 vessels reached the summit in 1795, 723 in 1800; 21,807 tons passed through Brimscombe Port in the former year, 41,698 in the latter. But far too great a proportion flowed in one direction, and that was not the direction, symbolically expressed in the design of the company's seal. Father Thames, far from showering his bounty on Madam Sabrina, drew heavily on her wealth. He supplied luxuries such as "1 Pipe Wine in a Case" and "4 Firkins Colour" brought down by a London boat and landed at Thames Head wharf for Thomas Estcourt of Estcourt House near Tetbury, but all the coal and more than three-fifths of the general merchandise moved eastwards, and the westward flow was little more than 17 per cent of the whole. Yet, although the return traffic was only enough to fill one in every two or three of the barges that reached London, the company were most unwilling to let any come back empty or lightly laden because the Thames Commissioners charged every barge per trip, out and home, whether fully loaded or not. Consequently numbers of the company's barges were always to be found waiting vainly at Hambro Wharf, while goods for shipment eastward piled up on the quays at Brimscombe and traders fumed at the delay.

Traders had other grounds for complaint. Delays arose from natural causes, flood or frost immobilising river or canal carrying. Consignments missed the connection at Brimscombe when the Severn tides obliged the Bristol or Stourport trows to leave before the Thames barges had arrived from London. Goods were frequently pilfered. From time immemorial dishonest transport workers have succumbed to the temptation to pilfer, but river boatmen seem to have been worse than most.[24] Thames bargemen were involved,[25] but not to the same extent as Trent boatmen whose reputation was so bad that Hugh Henshall & Co. felt obliged to put a 'supercargo' aboard every boat as it passed from the Trent & Mersey Canal to the river.[26] Severn trowmen developed it to a fine art, so often mishandling flour and grain on the company's Bristol trows that traders willingly paid more to send by road.

Frequently they ruined whole consignments of sugar or currants through adding sand and earth to make good the weight of what they had stolen. Furthermore, there was not much to be gained by claiming compensation, for the Thames & Severn company refused to be liable for loss or damage by fire, accident or leakage, unless caused by negligence — which would have been very difficult to prove; they would not accept responsibility for breakage of fragile goods such as glass or china; they would not carry wine in casks unless the casks were cased, and even then repudiated liability for pilfering unless the sender paid double freight; and on no terms whatever would they be responsible for the safety of wine in bottles.

Dissatisfaction came to a head at the end of 1801 when the lull in the war with France made the English Channel safer for the coasting trade. Even while the war had been on, Samuel Skey had proved that a coaster sailing direct between Bewdley and London could make more trips in a year than a barge could by canal, and now that the war was temporarily at an end, coastwise freights began to fall to their normal level, which was lower than by canal.[27] The company's agent at Brimscombe soon heard tell that west country traders were threatening to send goods by sea. Fortunately it was just at this juncture that the Thames Commissioners listened to the canal company's repeated requests that they should make some concession to barges returning from London without full cargoes. Moved, no doubt, by the company's claim that the canal trade paid nearly one-third of the total income derived from river tolls, the commissioners agreed to refund half toll to an empty barge and quarter toll to one half laden. Thereafter the Thames barges generally returned as they set out, in pairs, but with one laden and the other empty, and the London service was greatly improved. In little more than a year, too, Napoleon himself helped to reduce the advantage of coaster over canal barge.

Organisation of the company's scattered carrying and trading was the unenviable task of the superintendent of trade, Joseph Sills. He was a diligent and faithful officer, touring the canal, the Thames and the Severn, writing long reports covering everything of importance that he saw or heard, giving evidence before a parliamentary committee on the coal trade, negotiating on the proprietors' behalf with the Thames Commissioners. For a time, he even acted as clerk and kept the minutes. As his work increased, his salary was raised from £100 a year to £160, and he was allowed in addition £100 a year to employ a bookkeeper, £25 for a young clerk and another £25 for housekeeping and entertaining. Furthermore, the proprietors rewarded "his zeal and attention to their general concerns" by making him a gift of twenty guineas and an inscribed silver cup.

His particular concern was the coal trade. For more than twenty-five years the mainstay of the trade was Staffordshire coal drawn from Bilston. It came

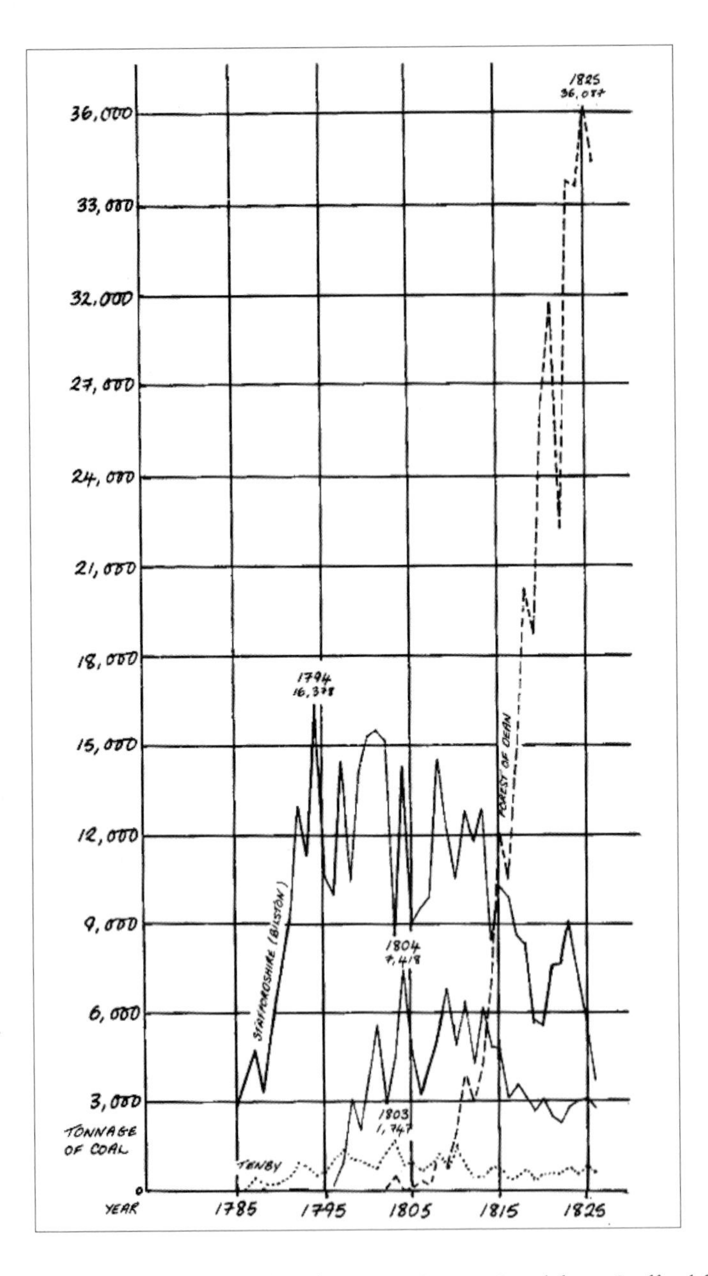

Tonnage of coal imported by the Thames & Severn Canal from Staffordshire, Newport, Tenby and the Forest of Dean, 1785-1826.

in narrow boats along the Birmingham and Staffordshire & Worcestershire canals to Stourport, where it was transhipped into trows which carried it to Brimscombe for storage on the island until it was sold locally or sent further east in Thames barges. The import began with 2,873 tons in 1785, exceeded 10,000 tons a year for a decade, and reached a peak of 16,378 in 1794. But Bilston was not an entirely satisfactory source. The state of the Severn often interrupted supplies. It was too long a haul and the cost of carriage and double transhipment raised the selling price uncomfortably. So although Staffordshire coal found a ready market in the Stroud valley, on the Cotswolds it could be undersold by coal brought 24 miles or more from Kingswood Chase by carriers who took back grain diverted from the company's Bristol trows. In 1798, the price in Cirencester of Sodbury coal (as they called it) was 19s. 6d. per ton delivered to the house as against 21s. per ton for Staffordshire coal on the canal wharf, and some twenty times as much was being sold. On the Thames, coal carried from the midlands in a single bottom via the Oxford Canal sold as far up as Eynsham Wharf and was even carted to Burford and Witney until the Thames & Severn offered a bounty of 2s. a ton on Staffordshire coal carried from Lechlade and simultaneously lowered their rates on grain for Bristol.[28] So it was evident that Staffordshire coal would never enable the company to achieve their aim of supplying the upper Thames valley with fuel or to fulfil their hope of entering the Oxford market, a hope the Oxford Canal Company were determined to thwart by fair means or foul. The latter indeed attempted to get through Parliament a clause prohibiting any coal which had passed along reaches of the Thames from entering their canal or being sold from their wharves, and when they were foiled in that, tried inducements, threats and even violence against traders attempting to sell coal brought via the Thames & Severn.[29]

It was imperative that the company should find a nearer source of supply than Bilston. Geographically the Forest of Dean was the most convenient, and from there they drew some coal experimentally in 1796-97, but the Forest coalfield was then undeveloped, transport to the water's edge was difficult, and when the coal proved to be of poor quality, no more was taken. Instead, the proprietors turned hopefully to a fast-developing coalfield which, although further away than the Forest, was nearer and far more accessible than Staffordshire. This was the eastern side of the South Wales coalfield, where the Monmouthshire Canal Company, formed in 1792, were building railways and canals linking the mines and ironworks with Newport, then a dingy little town recognised only as a 'creek' of the port of Cardiff. The canal works were completed in 1799, but much was already in working order when James Black visited the district in May 1796. Impressed with its great potentiality, he urged that the Thames & Severn company, who were already shipping some coal from Newport, should engage in the trade extensively.

This could not be done without new boats, as only the schooner *Swansea* was capable of trading with lower Severn ports and she was fully engaged in fetching from west Wales copper and special coal used in malting. Nor could it be done without reducing costs, as freight and tonnage increased the price of Monmouthshire coal from 9s. a ton in Newport to 20s. in Cirencester and 23s. 3d. in Cricklade, well above the price of the Sodbury coal it was hoped to displace. Believing that establishment of the trade would lead to a reduction of costs all round, Joseph Sills proposed a temporary sacrifice of 1s. a ton by the mines, 1s. a ton by the Monmouthshire Canal, and 2s. a ton (nearly half their toll to places east of Siddington) by the Thames & Severn.[30] He was so far successful that in 1799 Newport coal sold in Cirencester and Cricklade for 18s. and 20s. respectively, and imports to the canal, beginning with 201 tons in 1796, rose to 3,096 in 1798.

Trade with Newport brought to a head a dispute with the Customs, whose officers claimed that vessels sailing between Framilode and ports in the Bristol Channel were engaged in coastwise trade 'by open sea', and that their cargoes were therefore liable to duty under the statute of Charles II taxing sea-borne, as opposed to river-borne, coal. As it was, the captain of each of the company's Bristol trows had been obliged to make his way to and from Gloucester Customs House to obtain clearance before he could leave Framilode in either direction. But the liability of Newport coal to pay duty was disputed by the company on the ground that no part of the River Severn ever had been, or ever could be, legally considered 'open sea'. As the company refused to pay, the Customs detained more than one of their vessels early in 1796, and, in particular, impounded a cargo of beans and seeds, which they released only when the company had given a bond. Then, in order to settle the legal issue, a claim to the beans was lodged by the Crown in the Court of Exchequer. The case was tried by a jury, who found for the company, but the authorities applied for a fresh trial, which, after several days of learned counsels' arguments in February 1797 was granted. Thereupon, the Thames & Severn company, "finding that the determination of the Court of Exchequer was against them", submitted to the claim, and the Customs having gained their point, the bond was returned.[31]

In fact a reasonable solution of the dispute was at hand, for later in the same year a clause in an Act obtained by the Monmouthshire company granted exemption from duty on coal carried on the River Severn.[32] Moreover the Thames & Severn company persuaded the Treasury to consider stationing a Customs officer near the canal so as to avoid the vexatious delays to which their captains had been subjected, and the Gloucester Customs Collector, having assured the company that there would be no petty interference with their trade in future, expressed "his wish that the old grievance of the seizure of the seeds might be forgotten".[33]

The design of new vessels for the Brimscombe-Newport trade led to remarkable development of the Severn trow. The Schooner *Swansea* had not solved the problem of carrying cargoes to Brimscombe without transhipment as, in spite of her sliding keel, she was unable to work up the canals. Before ordering new vessels, the trading committee therefore experimented, raising the sides of existing trows by two feet or more, examining the relative merits of keels and lee-boards, and testing varieties of rig. Little time was lost: between the autumn of 1796 and the summer of 1797, the freeboard of *Loveden*, *Disney* and *Chalié* was increased, *Disney* and *Chalié* were fitted with keels, and *Chalié* rigged as a schooner with masts and sails as light as possible so that they could be lowered "with the greatest facility to enable her to pass thro' Newport Bridge".

Valuable experience was gained. *Loveden*, though considered fit for the Newport voyage, was actually confined to the Bristol trade, and, after plying to South Wales for some time, *Disney* joined her. *Chalié*, with a fore-and-aft sail on the mainmast, a lateen sail on the mizzen, and topmasts on both, proved cumbersome, and drawing 7½ feet laden, was as incapable as *Swansea* of passing up the canals, and therefore obliged to tranship her cargo to trows at Gatcombe or Newnham. The committee concluded that what they needed was a trow similar to the *Chalié* but rigged as a ketch. Five new trows were then ordered, four from The Bourn Yard and the fifth from Stourport, the programme financed by six proprietors each lending £200. The committee of trade optimistically expected the vessels to be ready in seven months. Under favourable conditions, each might perhaps have been launched in less than that and fitted out in a few months more,[34] but conditions at The Bourn were far from favourable.

In the first place, Bird's stock of timber was wholly insufficient: having ten trees, no more than enough for beginning one trow, he had to buy crooked oak and English ash, elm and fir, and, once the existing stock was exhausted, to use unseasoned timber "almost green to the great detriment of the Vessels so built". In the second, his labour force was neither large enough nor skilled enough for the rapid completion of such a programme. In the summer of 1799, there were two journeymen and four apprentices, paid from 1s. to 2s. 8d. a day, and even at its peak the staff never exceeded ten, including boatbuilders, smiths and a ropemaker. Their management was left far too much to Bird himself, who was allowed to take as many apprentices as he liked, and, as 'Inspector' of the yard, to pocket a subsidy of fourpence a day from the wages of each employee, with the result that he prospered at their expense. Further, when summer came and the men worked the longer hours customary in all trades at that date, the company failed to increase their daily wage, paying instead 'Extra Time' on the basis of an additional day or day-and-a-half a week, which of course benefited Bird (who suggested it) more than it did the men. Discontent was rife, and, as any skilled boatbuilder could find employment at 3s. a day in the busy yards at Chepstow and Newnham, "none but the most needy" would take service with the company. Even when summer wages at The Bourn were increased to 3s. a day, the staff remained dissatisfied and a change in Bird's position became essential. Early in 1800 he was obliged to accept a fixed salary of £80 a year, and as this was about £35 more than the yield of his previous daily wage, it is an indication of how much he had been pocketing. He was also ordered to stop making deductions from his men's pay, and to limit the number of his apprentices to two or three. About the same time, William Bird joined the staff at The Bourn on a similar salary.

Now, what had this small staff, more than half of whom were apprentices, to do? Unless itinerant sawyers were hired or manual labour engaged for some of the rougher work — and there is nothing to suggest that this was done — Bird and his men had themselves to man the saw pit, form the moulds according to the plan, lay the keel, make the frames, plank the hull, and lay the decks.[35] Then the boat was launched and the fitting out began. A ropeyard would be established, supplies of cordage, sail-cloth and blocks laid in, and the men would shape and step the masts, set up the rigging, prepare the yards, and make the sails. After that, they would caulk the decks, fit the ceiling or lining to the hold ("silling the hole" was Bird's rendering), fit up the cabin, and build the tow-boat. The tow-boat itself was quite a considerable piece of work: one of them was 23 feet long, 7 feet 4 inches wide, drew 2 feet 8 inches, and cost £21.

However, building new trows was not the only work of the yard, for the company had nearly fifty vessels, and Bird and his men were much occupied with repairs and even salvage. Of five boats sunk, three in the Thames and two in the Severn, four returned to normal service, but a trow that took Bird 8½ days to raise from the Severn reappeared as a lighter. For examination and repair of any badly damaged boat, there was only one dry dock, so that as pressure of work increased, each vessel had to be refloated at the earliest possible moment, often before it had properly dried. Moreover, this one dock was roofed so that work could continue in bad weather, but the roof was too low to admit the new trows. A second dry dock was therefore made in 1799. Each of the two was closed by a gate turning on bottom hinges, so that when opened it lay flat on the bed of the canal, and each was emptied by gravity through a conduit. Their overgrown basins can still be seen, and are now the only indication of the once busy boatyard at The Bourn.

Delayed by shortage of materials and skilled labour and the prior claims of maintenance and repair, the building of the new trows took between twenty-four and twenty-eight months — about four times the trading committee's estimate. The first, named *Fisher*, was ready in July 1799, but it is clear that she was built with unseasoned timber and indifferent labour because she was repaired on four separate occasions in her first fourteen months of service. She was followed by *Shadwell*, *Lane*, *Matthew* (built at Stourport but completed at The Bourn) and lastly *Newport*, launched at the end of August 1799 and fitted out in nine months more. *Fisher* and *Newport* were of 63 tons register, 63 feet 3 inches long, 15 feet 3 inches wide, with a hold 6 feet deep. *Shadwell* was slightly larger.[36] Four, *Fisher*, *Shadwell*, *Matthew* and *Newport*, were rigged as ketches, *Lane* as a schooner. The cost of the boats is not recorded, but a similar trow laid down later was sold, probably at the time of launching and while yet unrigged, for £280, and the schooner *Lane* fetched £393 2s. 8d. when less than four years old.

The design and construction of these boats show that the canal company were pioneers in the evolution of the Severn trow from a river vessel with square rig to a coaster with fore-and-aft rig. Twenty years later, trows were still normally square rigged,[37] and, it has been believed hitherto that the adoption of fore-and-aft rig, drop keels and false keels, in order to extend the range of the trow, was a development of the mid-nineteenth century.[38]

The schooners *Chalié* and *Lane* each carried 90 tons of coal from Newport, and in favourable conditions could make two voyages a fortnight, on the spring tide to Newnham and on the neap to Gatcombe where they were met by lighters, but such frequency could not be counted on. The ketches were handier to work and had the great advantage of being able to navigate the canals. They could carry, 68 to 80 tons from Newport to Framilode and 45 to 50 thence to Brimscombe, but as "it required from one to three tides in common times to go from Nuneham to Framilode on Account of the Shoals", and about a day-and-a-half more to reach Brimscombe, they could make only one voyage each spring tide.

The company's Severn craft were manned and worked in several different ways. Those trading up river to Stourport carried, like the Thames barges, a crew of three, and there is evidence that their masters were paid per voyage, as was common in the coasting trade[39] and customary on the Thames. Trows in the lower Severn trade were manned by five men and a boy, whereas the ketches and schooners could be handled by three and a boy. *Swansea* was worked at the company's expense by a crew paid regular wages: £3 a month to each man and 2s. 8d. a day to the captain, who was also given an allowance of a shilling a day per head (increased to 1s. 3d. at the close of the century) for victualling. The schooner *Chalié*, however, was worked in a way somewhat similar to coasters sailing from some south coast ports earlier in the century, at the expense of her master, who charged freights of 3s. and 3s. 6d. a ton from Newport to Gatcombe and Newnham respectively and paid the company one-third of this for the use of the vessel.

Having equipped themselves with six boats suitable for the lower Severn trade and knowing there were great reserves of good coal in the neighbourhood of Newport, the company had reason for confidence that a satisfactory trade would develop. There was even discussion of sending Newport coal through the canal to London, where the rising cost of fuel caused such concern that in 1800 a parliamentary committee examined the possibility of using the inland navigations to break the virtual monopoly of the London market held by the Tyne trade. Joseph Sills was one of the witnesses examined, and expressed the opinion that if production were stimulated, 50,000 to 100,000 tons a year might be sent to London. However, neither the quantity nor the price, inflated by freights, tolls and duties to 66s. 3d. a London chaldron of 27 cwt, would have had any effect in reducing the domination of the Tyne trade, which in 1800 supplied over a million chaldrons at a price of 60s. to 70s.[40]

It is doubtful whether the production of the Newport mines could have been quickly stimulated to the extent suggested by Sills. As it was, the output was insufficient to meet the local consumption of the furnaces, forges and rolling mills as well as the export trade. Visiting Newport early in 1800, Sills learnt that twenty boats from Bridgwater, Plymouth and Ireland had been unable to obtain cargoes the preceding week, and in 1801 the Thames & Severn company found that their two schooners and four ketches, although capable of importing nearly a thousand tons a month, brought no more than 5,545 tons in the year. Moreover, the Monmouthshire coal proved disappointing not only in quantity but also in quality.

So far, therefore, from Monmouthshire coal becoming the principal supply distributed along the canal, the quantity handled was rarely more than half that of Bilston coal. The import of the latter reached its peak before the Newport trade began, but in each of the three years 1800-1802 it exceeded 15,000. The five trows and two frigates engaged in the Stourport trade therefore managed to bring about 1,250 tons of coal a month, in spite of the unsatisfactory state of the Severn. There was talk of increasing this by employing a new type of vessel called a "Severn Flat", able to voyage however low the water. One of these, carrying 20 tons on a draught of 18 inches, double the load of a trow under similar conditions, had been seen at Brimscombe in October 1797, but although the committee of trade authorised Perry and Black to obtain two, there is no evidence that they succeeded in doing so.

Besides the trows engaged in the import of coal, there were two in the Bristol trade, *Disney* and *Loveden*, worked alternately by the same captain and crew. These sufficed until the rates for grain for Cirencester were reduced, when the trade increased so much that a third boat became necessary. This, a trow similar to *Disney*, was ordered from The Bourn Yard in January 1800, and when completed in August 1802, was 64 feet long, 15 feet 4 inches wide, with a draught of 5 feet 7 inches and a burden of 85 tons.

By the beginning of 1800, the Thames trade also exceeded the capacity of the company's boats, although they had thirty-one Thames barges. Four more were added in 1800-1802, one of which, *Commerce*, built at The Bourn for £150 in 1801, differed significantly from her predecessors: 64 feet long and 12 feet 4 inches wide, with a capacity of 46 tons, she was of similar size to the *Droitwich*, short enough for the locks below Brimscombe and narrow enough for those above, and therefore able to do what James Black had advocated years before — act as a lighter to the Welsh vessels in the Severn and carry the coal direct to the Thames.

Hitherto, the regular use of such a barge, other than those engaged in the salt trade, had been prohibited during the season of short water. This meant that any private carrier who wished to trade regularly along the whole length of the canal had to invest in two different types of vessel for service east and west of Brimscombe, and it is probably the main reason why so few of them. used the canal at all (Black thought there were only about four in 1796). Even when plentiful supplies of water rendered the prohibition inoperative, very few carriers availed themselves of the opportunity, and indeed it is said that the first ever to do so was "Mr Yates, master and proprietor of a canal barge at Coalbrook Dale", who in July 1800 "went all the way, which is upwards of 400 miles by water, from that navigation to Hambro' Wharf, near London Bridge, in 14 days".[41]

Until some encouragement was given to private traders, the company were therefore obliged to continue carrying and trading. While they did so, they employed at least 250 persons. About 170 of these manned the fleet of more than fifty vessels, 64 were in regular employment in the offices, agencies, workshops, watch-houses and boat-building yard, and unknown numbers were employed in towing and on the quays, where (as the records mention few port workers at Brimscombe and none at all at Stroud, Cirencester, Cricklade or Lechlade) casual labour was probably engaged as required. In addition, navvies were occasionally hired to assist the permanent staff in the execution of major repairs. The committee's claim that they were "a large Company" — it was a lamentation, not a boast — was justified, for by contemporary standards they were. At that date there were very few large concerns, and even thirty years later some of those which ranked as such had less than a hundred employees.[42] The proprietors of the Thames & Severn, becoming aware after a time that "a large Company [was] necessarily exposed to impositions, which Individuals may avoid", were driven to the conclusion that efficient management of "so complicated and distant a concern, as the trade between London and the Ports on the River Severn" was beyond the capacity of their simple organisation.

AN ESTIMATE OF THE STAFF EMPLOYED WHILE THE COMPANY WERE CARRYING AND TRADING

Superintendent of trade in London

Clerk-of-the-works at Cirencester or Siddington

Cashier and navigation agent at Brimscombe Port — posts combined 3 October 1798

London office, staff of three:

housekeeper
clerk or book-keeper
boy

Brimscombe Port, staff of about seventeen:

2 clerks
2 apprentices
housekeeper
servant maid
4 wharfmen
3 or more labourers
1, or probably 2, carpenters
mason
blacksmith

Thames Head, staff of four:

carpenter
carpenter's mate
engineman
fireman

Agents along the line of the canal, five:

Walbridge, Stroud
Siddington
Cricklade (Latton),
Kempsford
Lechlade

Agents off the line of the canal, seven:

Bristol
Gloucester
Tewkesbury
Upton-on-Severn
Worcester
Stourport
Abingdon (a clerk)

Watchmen along the line of the canal, twelve;

Stroud
Brimscombe
Chalford
Furzen Leaze
Cirencester
South Cerney

Puck mill	Cerney Wick
Sapperton	Marston Meysey
Coates Field	Inglesham

Two carters working between Lechlade and Abingdon

The Bourn Yard, staff of ten:

inspector	blacksmith
2 journeymen	blacksmith's assistant
4 apprentices	rope-maker

Men for the fleet of trows and barges, about 171:

12 cost bearers of Thames barges
99 for crew of 3 men each on 31 to 35 Thames barges
5 for the crew manning two Bristol trows
12 for the crew of 3 men and a boy on each of the three schooners
16 for the crew of 3 men and a boy on each of the 4 ketches
15 for the crew of 3 men on each of the 5 Stourport trows
6 for the two frigates, having probably 3 men each
6 for 3 lighters, say 2 men each

Total 235

Men for towing along the canal, number unknown

Men for port work at Stroud, Cirencester, Cricklade and Lechlade were probably recruited from casual labour as required, and the Brimscombe staff augmented in the same way

Navvies for the annual overhaul and repair were recruited from casual labour at a later date, and may have been occasionally in the earlier years

REFERENCES

1. See, for example: Jackman, W. T:. *The Development of Transportation in Modern England*, 1916, vol. 1, pp. 434-6; Cadbury, G., & Dobbs, S. P., *Canals and Inland Waterways*, 1929, pp. 20-1;. Willan, op cit, 1936, p. 115; Hadfield, C., *British Canals*, 1950, pp. 64-5.
2. Halsbury's *Laws of England*, second ed, vol. 27, p. 336, section 757.
3. Act 8 & 9 Vict. cap 42.
4. Act 23 Geo. III cap 38 section 57. Similar entries occur in, for example, the Acts of the Stourbridge and Dudley canals.
5. Jackman, op cit, vol. 1, p. 435.
6. Browne's *Bristol Directory*, 1785, p. 75.
7. Kippis, op cit, vo1. 2, p. 601, note quoting an advertisement dated 1 July 1777; also advertisement in GJ 20 February 1786.
8. Hadfield, C., *British Canals*, pp. 149-150.
9. Jackman, op cit. vol. 1, pp. 435-6.
10. Herefordshire & Gloucestershire Canal Report of the Committee 10 November 1796 (GCL).
11. Birmingham Canal, Minutes of the Committee to conduct the Coal Trade (BTR).
12. Abstract and General 'State of the Accounts of . . . the Stroudwater Navigation', April-October 1819.
13. Stourbridge Navigation Minute Book, 7 July 1783 and 3 July 1786 (BTR).
14. *Observations on the general comparative merits of Inland Communication by Navigations or Rail-Roads, by a proprietor of the Kennet & Avon Canal*, 1825.
15. Hadfield, *British Canals*, p. 65, quoting from the Basingstoke Company's report of 13 February 1806.
16. I have not succeeded in tracing any such action, nor reference to one in the parliamentary proceedings of 1845, but the proprietors of the Upper Medway Navigation were certainly threatened with one (see Hadfield, C., *Canals of Southern England*, pp. 106-7).
17. See *Reports of the Engineers . . .* 1791, pp. 50-1.
18. Grant, J., *The Narrative of a Voyage of Discovery . . . To which is prefixed, An Account of the origin of sliding keels*, 1803, p. xxiv and see also pp. v to xxvi. Steel, D., *The Elements and Practice of Naval Architecture*, third ed. 1822, pp. vii, 159-162 and plate 19.
19. See Commissioner, op cit, 1767, p. 15.
20. Sutcliffe, op cit, p. 237.
21. Rees, *Cyclopaedia*, Canal.
22. Atkins, J., *The Tradesmen's Tokens of the Eighteenth Century*, 1892, p. 33. See also Thorburn, W. S., *A Guide to the History and Valuation of the Coins*

of Great Britain and Ireland, 1905, pp. 93-4; Ashton, T. S., *The Industrial Revolution 1760-1830*, 1948, pp. 99-100; Clapham, op cit, p. 265.

23. PP 1793 Reports 13/109, p. 32, evidence of James Black.
24. See Jackman, op cit, vol. 1, pp. 360, 440-1; Rees, *Cyclopaedia*, Canal; Sutcliffe, op cit, p. 140.
25. See Willan, op cit, pp. 108-9, an example of 1632, and Thacker, op cit, vol. 2, p. 444, an example of 1829.
26. Kippis, op cit, vol. 2, p. 601 note.
27. Sutcliffe, op cit, p. 376, and *Observations on the general comparative merits* . . . Both state that the expense of carriage was less, by coaster than through the Kennet & Avon Canal.
28. PP 1793 Reports 13/109, p. 25, evidence of T. Court. Mylne, R., *Report . . . on the present state of the Navigation of the River Thames, between Maple-Derham and Lechlade*, 1802, p. 33. Letter from J. Sills to — Boucher 25 April 1798 (BTR). Various references in the T & S papers. See also Rudder, S., *The History of the Ancient Town of Cirencester*, 1800, p. 162, for a complaint that coals had not been "rendered cheaper to the town than they were when brought by land from Gloucester".
29. See Oxford Canal Bill 1798, and Reasons against the Bill January 1799 (Oxford City Library P.328/37); and Memorandum, Oxford Canal, exclusion of coal from the Thames and Severn Canal (GCRO).
30. Letter from J. Sills to — Boucher 25 April 1798 (BTR).
31. PP 1810 Reports vol. 4, p. 226, and see also pp. 151-238; also T & S Collection.
32. PP 1810 Reports vol. 4, p. 226. Priestley, op cit, pp. 485-7.
33. Report of J. Sills, 9 January, 1800 and other T & S records.
34. Cf Greenhill, B. *The Merchant Schooners*, 1951, vo1. 1, for details of the building of wooden ships and the time taken to fit out, especially pp. 41, 55, 62.
35. Ibid for excellent descriptions of the work of a small yard, particularly pp. 59, 74 and also 41, 54-5.
36. Registrations of the Port of Gloucester, Gloucester Customs House.
37. Dupin, op cit, vol. 2, p. 333.
38. *The Mariner's Mirror*, Farr in vol. 32, pp. 85-6, Nance in vol. 2, p. 201, Greenhill in vol. 26, p. 286.
39. Cf Willan, T. S., *The English Coasting Trade 1600-1750*, 1938, pp. 17-8, 20.
40. PP 1800 Reports from Committees on the Coal Trade vol. 10, pp. 567, 645, evidence of Joseph Sills, and also pp. 538, 643. Cf Ashton & Sykes, op cit, pp. 248, 251, 253.
41. Phillips, op cit, p. 592.
42. Clapham, op cit, pp. 68, 69, 154-5, 165, 176, 185, 186, 191.

CHAPTER NINE

MONEY AND MEN

Every canal company hoped that the share capital would suffice to equip the line and, in addition, provide a dividend of 5 per cent per annum until revenue should be earned. This payment of interest out of capital was common practice in transport undertakings until 1847.[1] It encouraged proprietors to subscribe promptly and the committee to see to the works being completed as quickly and economically as possible, but it could also result in false economy. Baron Dupin, criticising the inferior workmanship to be seen on some canals, remarked that it arose "sometimes from negligence, often from ignorance, but more generally from a motive of economy".[2] It was certainly the reason for the narrow bore of Harecastle tunnel which so disastrously restricted the size of the standard British canal boat, and for that and other tunnels devoid of towing-paths.

Even so, miscalculation of estimates, alterations of plan, rising prices, frequently rendered the original share capital inadequate, so that more money had to be raised, either by private subscription as the Stroudwater proprietors did, or by issuing more shares or raising loans.[3] In fact most Acts authorised loans for use if necessary, by the issue of mortgages, bonds or promissory notes, all bearing interest at 5 per cent. As until 1794 mortgages bore no time limit,[4] and as bonds and notes were often issued on the understanding that they could be converted into shares, it seems certain that the lenders looked on them more as long-term fixed-interest investments than loans, just as later generations looked on preference shares. Furthermore, many of the lenders were original shareholders anxious to see the works completed; and so in case of default it was not difficult for debtors and creditors to conclude some mutually satisfactory arrangement,[5] more especially as foreclosure offered no advantage, canal property having little value unless the navigation remained in use.

Before the Thames & Severn Canal was finished, not only was the share capital used up, but also the £60,000 authorised to be borrowed on mortgage. Power to borrow another £60,000 on similar terms to complete the equipment of the canal was therefore obtained in 1791. Meantime it was not of course possible to pay interest in full out of capital, so although holders of less than five shares received cash for two years' interest in 1788, the rest were fobbed off with interest bonds.

There was a muddle in issuing these, as mortgage forms were 'inadvertently' distributed and had to be called in. Further, it is clear that some unauthorised person had affixed the company's seal to them, as it was decreed that henceforth the seal was to be kept in a box with three separate locks and only used at properly constituted meetings. The incident was a symptom of the company's thoroughly inefficient management at that time. There was no strong guiding hand. It is a curious feature of the Thames & Severn and other early canal Acts that they make no mention of a chairman.[6] The omission was probably deliberate to avoid any one person acquiring undue influence. Yet someone had to preside each time the proprietors met. At meetings of the Stroudwater, the selection is said to have been made in a most entertaining way:[7] the company were bidden to assemble at a certain hour, punctuality was de rigeur, and the last proprietor to enter the room on the stroke of the clock took the chair. Perhaps there was sometimes a nice show of politeness on the threshold of the room, a becoming reluctance to enter before a colleague! But whether a choice was made in this way or, as later Acts stipulated, by election, it is clear enough that the chair need not have been occupied regularly by the same proprietor. The minute book of the Thames & Severn, for example, records the chairman's name on only three occasions prior to 1800 and each time it is different.

In the absence of a permanent chairman, great responsibility devolved on the treasurer, but it so happened that three different bankers held this office in succession in the years before 1800. As for the clerk, the salaried servant who might have been expected to prevent such an error as the 'inadvertent' issue of the mortgage forms, there was none for nearly twenty-nine years after Joseph Grazebrook resigned. Meantime some of the duties were covered by two solicitors who in succession handled legal matters, others by three successive officers who kept (and often signed) the minutes. Very few proprietors troubled to keep a watch on affairs. Twenty-two were present at their first general assembly held at The London Tavern, but attendance dropped to five or ten at succeeding meetings held in Cotswold inns, and when the assembly room at Brimscombe Port was used, two meetings out of nine were adjourned for lack of a quorum. Later assemblies were almost all held in London, followed by dinner in a tavern, but even then an attendance of eighteen was good.

What little direction there was came from the thirteen members of the managing committee, who met frequently and once a year made a thorough inspection of the canal in their private barge. The inspection was a leisurely progress with nights spent at Brimscombe, Cirencester and Lechlade, and was obviously an enjoyable occasion, even if the food was less lavish than in London, where their menu included salmon, chickens, goose, lamb, sweetbreads, pudding, tart, washed down with beer, sherry, Madeira and port, all for as little as £4 or £5 for the lot of them.

John Stevenson Salt, treasurer, 1800-1845, as a young man.

John Disney, first permanent chairman, 1805-1828.

Although the shareholders were without dividends, they were not yet without hope, as it was still believed that expansion of the company's carrying and trading would bring better times; and shares actually sold at a premium in the first quarter of 1793, fetching 134½ in January and 115 in March. But as 1795 drew to a close and there was little evidence of real improvement, the committee, still anxious to invest every penny carrying, asked the mortgagees to forego their interest and the shareholders to approve an ingenious, if not wholly creditable, plan for capitalisation of their outstanding interest. This by then amounted to £37 10s. per share, and the idea was that the holder of each original share should pay £12 10s. in cash and receive a half share in exchange. Many, James Perry amongst them, vigorously opposed the plan, resenting what they felt was an attempt to coerce them into paying cash in order to preserve their rights. Nevertheless, the half-share plan was authorised by Act of 1796, which, however, preserved the right of dissentients. Finally, a few shareholders took up more than their allotment, and oddly enough some half-shares were purchased by newcomers, but 426 of the 1,300 remained unsubscribed.

Widespread promotion of canals during the 'mania' of the 1790s improved the prospects of the Thames & Severn a little. Certainly one, the Kennet & Avon, was bound to take away their London and Bristol trade, but it was expected that four others — the Herefordshire & Gloucestershire, Worcester & Birmingham, Leominster and Gloucester & Berkeley — would increase traffic to London via Stroud.[8] Best of all, formation of a sixth, the Wilts & Berks, offered an opportunity of avoiding the Thames navigation above Abingdon, for near Cricklade the two lines were close and a junction could readily be made. But it was to the Herefordshire & Gloucestershire that proprietors of the Thames & Severn subscribed most freely, taking up twenty shares between them.[9] Amongst those who did so were Chambers and the Chaliés, who, together with Littleton and the Lanes, were prominent members of a group whose readiness to lend money privately without the protection of a mortgage can only be explained by their investments in other concerns. In all, £11,000 were raised by private loans in twelve years. Some were short-term, such as those for building boats which remained the lender's property until six annual instalments had repaid principal with interest at a little more than 12½ per cent. Some were for purchase of coal which was later sold. But one sum of £2,000 was lent for no specific purpose and on nothing more substantial than the lenders' "Faith in the Committee". Now of the seventeen proprietors who were parties to these private loans, thirteen — including all those who participated in more than one — stood to gain from increased sales of Staffordshire coal, possibly as mine-owners, certainly as shareholders in the Staffordshire & Worcestershire or Stroudwater canals.

But whether the company's carrying and trading was really profitable was being questioned by some proprietors as early as 1800; however even Joseph

Sills, the superintendent of trade, found it "impossible without immense Labour & difficulty to draw any accurate Conclusion of the profit or loss". So William Stevenson, who had resigned as treasurer six years earlier but who held £10,000 worth of mortgages, initiated discreet investigations, and in May 1800 his able young nephew and son-in-law, John Stevenson Salt,[10] who was already on the managing and trading committees, was appointed to the key position of treasurer at the age of twenty-four. A year later, a confidential clerk, one Harry Harford, was installed at Brimscombe, and very soon wrote telling Salt that it would be to the company's advantage if the carrying was in private hands and "if the Canal & its Trade had had justice done them by the different persons employ'd thereon". In fact, his meticulous investigations revealed "extravagant charges" for repairing boats; widespread pilfering of coal; failure to charge the committee of trade interest on the capital advanced; a deficiency of £50,874 between the balance of stock and the number of shares and half-shares; and the malpractices of employees — in particular the preoccupation of the Brimscombe agent with a private trading venture of his own and his fraudulent conversion of the company's property. This last piece of news brought Salt himself to Brimscombe, and when his own investigations disclosed irregularities in the handling of cash payments, the agent was relieved of his duties.

The temptation to which this man succumbed was very strong: The current system — if system it can be called — of accounting for cash payments was no doubt convenient at a time when communications were slow and hazardous, but it was open to grave abuse and was maintained far longer than was necessary. It was the practice for all the local agents to use the cash collected on their wharves for the payment of rates, taxes, wages and incidental expenses, and to account only for the balance, which was periodically remitted to head office and thence to the treasurer or bank. Accuracy therefore depended entirely on each man's integrity. In the twenties and thirties of the nineteenth century, the Brimscombe agent compiled a weekly cash account with care, remitting £5 and £10 notes for greater security in halves sent separately by letter and coach, but his less meticulous successor was dismissed for misappropriating more than £3,000. Later in the century even so respected an officer as John Taunton drew from the company's cash box in meeting personal expenses, and although the amounts were later deducted from his salary, his action threw the accounts into such confusion that in 1884 an accountant found it impossible "to effect a complete and proper audit" within the time at his disposal.[11]

The state of affairs disclosed by Harry Harford in 1801-1803 so shocked the managing committee and the proprietors that they reversed their previous policy and began to contract the carrying and trading. Between 1802 and 1805 the outlying agencies were closed, the boat-building yard at The Bourn was let to Samuel Bird, and boats were sold or leased to suitable operators. In the process, the permanent staff was reduced to a very small number,

AN ESTIMATE OF THE STAFF EMPLOYED AFTER THE COMPANY HAD ABANDONED CARRYING AND TRADING

Clerk-to-the-company, part time only
Agent or Manager at Brimscombe Port
Clerk-of-the-works at Siddington

Brimscombe Port:

3 clerks or apprentices (later only two)	3 or more labourers
housekeeper	2 carpenters
servant maid	blacksmith
4 wharfmen	mason

Thames Head:

engineman	fireman

Siddington:

2 carpenters

Agents along the line of the canal:

Thames Head Wharf until about 1835 only	Cricklade
	Lechlade

Watchmen along the line of the canal, twelve:

Stroud
Brimscombe
Chalford (later called lock-keeper)
Puck Mill
Sapperton (later called lock-keeper and stationed at Daneway)
Coates Field
Furzen Leaze or Blue House
Cirencester (later called wharfman)
South Cerney (later called lock-keeper)
Cerney Wick (later transferred to Wilmoreway)
Marston Mersey (later transferred to Eisey)
Inglesham (later called lock-keeper)

Total 38

probably no more than thirty-eight.

As the net earnings were no longer needed for expansion of the carrying, it became possible to pay something to the mortgagees, but when these had received only four payments of 2 per cent in four years, the more far-sighted of them realised that some arrangement would have to be made between shareholders and creditors. It was probably the realisation of the difficulties which would be involved that led them first to strengthen the company's organisation.

The most important step was the selection of a chairman whose appointment became permanent in fact if not by intent, for there is no comment in the minutes until the day of his resignation twenty-three years later. This was John Disney junior, a barrister and son of the Rev Dr John Disney, a distinguished divine and a wealthy man with many secular interests[12] who had himself often presided at the company's meetings. Holding 54 shares and £3,700 in mortgages between them, the Disneys were in a strong position to give a lead in negotiations between proprietors and mortgagees, and this position was strengthened when John Disney junior married his cousin, daughter of Lewis Disney-Ffytche who held 91 shares and £22,715 in mortgages, making him by far the largest of the company's creditors. After a few years as member of the trading and managing committees, the younger Disney first took the chair in 1805 at the age of twenty-five. He proved an excellent choice.

About the same time, another appointment of far-reaching consequence was made, that of a new agent at Brimscombe Port. This was John Denyer, a young man of twenty-four, son of a Freeman of the City of London and an ex-pupil of Christ's Hospital[13] who had joined the company in May 1794 as an apprentice in the Brimscombe office. He had made his mark even before serving his time, for he had taken the journal and ledger of the troublesome trade account to London for examination by the committee of trade, who had given him twenty guineas "in consideration of his commendable attention to the Books". He then had a spell as bookkeeper in the London office before returning to Brimscombe in 1804 as 'Clerk or Manager' at a salary, including allowances, of £230 a year. As the carrying and trading contracted and the influence of Joseph Sills declined, the post of Brimscombe agent became the most important in the company's service, and this Denyer held for more than thirty-eight years.

And so, early in 1805, the three principal offices of chairman, treasurer and manager were filled by men still in their twenties. Their partnership remained unbroken for twenty-three years and during that time the prospects and efficiency of the concern were transformed out of all recognition.

In tackling the problem of the debt, however, the older generation of proprietors — Littleton, Matthew Chalié and the Lanes among them — played at least as prominent part as Salt and the younger Disney. It took several years to evolve a solution. For one thing, it was not possible to estimate the company's income until the carrying business had been disposed of, and so it was 1808 before the committee were able to give proprietors and creditors a full analysis of the position. Their statement was discouraging enough: the principal debt consisted of the mortgages and bonds with arrears of interest, and when to this was added the interest due to those proprietors who had not subscribed to the issue of half-shares, the total amounted to "the enormous sum of £193,892 10s.". In the committee's opinion it was "absolutely impossible" that this debt could ever be redeemed, and most unlikely that earnings would even suffice to pay the interest on the mortgages and bonds, let alone provide anything for the shareholders. In their opinion, therefore, the only wise course was to relieve the company entirely of debt.

Their conclusion was widely approved, and as the members of the committee "and their immediate connections" were holders of nearly half the mortgages and bonds, they were able to proceed on the assumption that the aims of both proprietors and creditors were substantially the same, namely to give some satisfaction to the creditors without expropriating the shareholders. The real problem was how to evolve fair and equal treatment for those proprietors who had subscribed for the half-shares in 1796 and for those who had not done so, as the former had made a cash payment of £12 10s. which the latter had avoided, and had accepted capitalisation of £37 10s. interest to which the latter remained legally entitled. On this point the managing committee's initial plan failed to gain acceptance, and a special committee appointed to produce another found themselves so "delayed by the great complexity of the business" that it was nearly a year before they completed their task.

When they had done so, their plan showed remarkable foresight and generosity. The income of the company appeared sufficient to pay 1½ per cent per annum on the principal debt, and there were assets which could be realised to provide a substantial sum in cash. So the committee suggested that "the cheapest and best method of foreclosure" was for the creditors to accept a cash payment of 21 per cent in discharge of arrears of interest, and to exchange their securities for new shares of equal nominal value, entitled to 1½ per cent interest "secured . . . by a prior lien upon the Tolls of the Canal" in effect preference shares, but at that date there had been scarcely more than half-a-dozen examples of such preferential treatment and the name was not in general use.[14]

The plan met with general acceptance and was sanctioned by Parliament in 1809. 1,150 New or Red £100 shares were issued in exchange for the mortgages and bonds, and the 1,300 Old or Black shares were left intact. It was provided that if the revenues of the company should fall short of the sum required to pay the full interest of 1½ per cent on the Red shares, the right of the holders should not be cumulative; but if the revenues should exceed that sum, the excess was to be applied in distributing up to 1½ per cent on the Black shares, and thereafter in equal division between all. A neat solution of the problem of the half-shares was achieved by cancelling all those which had been issued, and by arranging that proprietors who had not subscribed for any should forfeit £12 10s. out of future dividends. An interest in the prosperity of the concern was therefore preserved for all, and in course of time, all, even the non-subscribers, benefited.

The reconstituted capital of the company, £245,000 in Red and Black shares, remained unchanged for sixty-seven years. It represented an investment of about £8,130 per mile. It is difficult to find comparable figures for other canals: while Joseph Priestley in 1831 recorded the amounts authorised in shares or on mortgage, he rarely stated what was in fact raised. Henry English in 1824-1825 recorded share capital but not loans.[15] Moreover, any attempt at comparison is affected by the width of the canal, by rising prices, and by the increasing scale of engineering. Available information suggests (it is no more than that) cost per mile of about £1,050 for the Bridgewater,[16] and investment per mile of £1,390 for the Trent & Mersey, £3,350 for the Chesterfield, £9,430 for the broad Leeds & Liverpool, £17,800 for the narrow Huddersfield and the broad Kennet & Avon.[17]

The redemption of the crushing debt and the influence of the young and energetic management infused, as the committee said, "new energy . . . into the whole system" and put "this once unpromising concern . . . upon a firm . . . footing"; so the shares soon rose in value. Whereas in 1805 original shares sold for as little as £3 or £5 and £250 interest bonds fetched only £52 or £54, by July 1808 news of the impending reconstruction caused ordinary shares to appreciate to £13 10s. or £14 10s., and by the spring of 1811, after dividends of 30s. had twice been paid on both Red and Black shares, Reds cost £32 each.

The improvement, however, was never enough to remove the Thames & Severn from the category of financially unsuccessful canals. It was one of many whose success must be measured, not in terms of shareholders' dividends,[18] but by the benefit they conferred on the communities living along the line. They all reduced the cost of carriage, made plentiful supplies of coal available in fuel-starved country districts, increased the value of local products of all kinds, and stimulated trade generally.[19] There is no reason to doubt that the merchants of London, Wolverhampton and Bristol, the Staffordshire coal-owners, and the proprietors of the Staffordshire & Worcestershire Canal had calculated, as Dupin said canal shareholders commonly did,[20] that the advantages they would derive from the use of the Thames & Severn would be 'far superior' to the dividends they might or might not receive from their investment in it.

REFERENCES

In the preparation of this chapter, I have been deeply indebted to Evans, G. H., *British Corporation Finance 1775-1850, A Study of Preference Shares* (Baltimore, John Hopkins Press, 1936).

1. Evans, op cit, pp. 2-4, 76-9, 119-120, 151. The Acts of the Trent & Mersey, Staffordshire & Worcestershire, Coventry, Birmingham, Oxford, Stourbridge, Dudley and Monmouthshire canals, for example, include this provision, but that of the Droitwich does not.
2. Dupin, op cit, vol. 1, p. 339.
3. Priestley, op cit, mentions at least forty examples of supplementary funds having to be raised in various ways. See also Evans, op cit, pp. 45-50, 53-6.
4. Evans, op cit, pp. 47, 54, 55.
5. Arrangements of this kind were authorised in Acts obtained by the Edinburgh & Glasgow Union and Oakham canals (Priestley, op cit, pp. 251, 522), as well as by the Thames & Severn.
6. There is no mention of a chairman in, for example, the Acts of the Trent & Mersey, Coventry, Droitwich, Oxford, Stourbridge and Dudley companies, but there is in those of the later Monmouthshire and North Wilts companies. See also Dupin, op cit, vol. 1, p. 136.
7. According to the late P. G. Snape, who heard it from his father W. J. Snape. The Snapes, father and son in succession, filled the post of clerk to the Stroudwater Navigation from 1872 to the 1950s.
8. PP 1793 Reports 13/109, p. 16, evidence of Josiah Clowes.
9. State of the Accounts of the Herefordshire & Gloucestershire Canal Company, 7 November 1796 (GCL).
10. In tracing the story of the Stevensons and Salts, I am indebted to the secretary of Lloyds Bank and the editor of *The Dark Horse*, the staff magazine. Other sources are: *Express & Star & Birmingham Evening Express* 15 June 1937, article by Quaestor; *The Dark Horse* 1930, vol. 11, pp. 542-3; Roxburgh, A. L. P., *Know your Town — Stafford*, 1948, pp. 78-80; article on Stafford Old Bank; Berkswich Women's Institute, *The Story of Berkswich*, 1949 (duplicated); *Burke's Peerage, Baronetage & Knightage*, Salt of Weeping Cross — John Stevenson Salt was born 25 June 1775.
11. Report of investigation made into the Thames & Severn Canal Co.'s Accounts by J. N. Mahon, GWR Accountant's Office, Paddington, 7 July 1884 (GCRO).
12. DNB John Disney DD 1746-1816 and John Disney 1779-1857.
13. I am indebted to the clerk of Christ's Hospital for the record of Denyer's admission to and discharge from the school, the date of his birth (18 April 1780), and his father's position.
14. See Evans, op cit, pp. 73-5, 164.

15. English, H., *A Complete View of the Joint Stock Companies*, 1827, pp. 35-7.
16. Anonymous, *The History of Inland Navigations*, 1766, p. 41, and see also p. 58 where it is stated that Brindley expected to build the Trent & Mersey Canal south of Harecastle tunnel for £700 a mile, and north of it for £1,000 a mile.
17. Priestley, op cit, pp. 162, 427, 682. English, op cit, p. 36.
18. The size of canal dividends has been greatly overrated. Sutcliffe (op cit, Preface p. ii) in 1816 deplored "the generally prevailing idea, that canals were extremely beneficial to the subscribers", and the *Quarterly Review* (vol. 32, p. 170) in 1825 contained a careful analysis of the published statements of eighty canal companies, which showed that the large dividends of 20 per cent or more were paid by no more than ten, and that the average overall was about 5¾ per cent.
19. See, for example, the remarks in the *Quarterly Review*, 1825, (vol. 32, p. 171), in Clapham (op cit, p. 80), and in Ashton & Sykes (op cit, p. 149) quoting from "A Treatise upon Coal Mines", 1769, the comment that labourers in country districts had often been unable to dry their clothes or have hot food after working in the rain.
20. See Dupin, op cit, vol. 1, pp. 116-7.

CHAPTER TEN

LINKS IN THE CHAIN

It was easier for the proprietors to decide on withdrawal from the carrying and trading than to achieve it, especially as they insisted that only 'proper persons' should be allowed to take part. Obviously it was easier to dispose of the boats designed for the Severn where trade was well established, than those intended for the trade from Brimscombe down the Thames which had caused the company such loss. The *Swansea* was sold to her master, Thomas Chapman;[1] the ketch *Newport* to a Chepstow owner;[2] the unfinished trow to Benjamin Grazebrook who named her *Stroud Galley* and used her in the Stroud and Bristol trade for which she had been designed. Christopher Bowly, whose family was well established in Cirencester,[3] bought the schooner *Lane* and a Thames barge, no doubt to supply the town with coal. But some who were allowed to lease vessels experimentally failed to make good, and although five carriers eventually formed an association reliable enough to take over the bulk of the trade, the Thames & Severn company still owned, even if they did not operate, twenty-five Thames barges, the largest fleet on the river west of London, as late as 1812.[4]

Joseph Sills claimed that the new carriers aimed at "a limited trade with a liberal profit",[5] but the evidence of the company's own experience and of the carriers' activities and subsequent history (so far as it appears) does not confirm this. Rather, it suggests that their profits were meagre, their resources easily strained by disaster, the margin between success and failure slender. Certainly the carriers gained two solid advantages: in 1804 they persuaded the canal to accept, in lieu of the variety of authorised tolls, a consolidated charge of 4s. 6d. a ton on all goods except coal and grain; and using the *Commerce* built in 1801, they were able to carry in one bottom from Stourport or the Severn estuary to the Thames. The importance of this development cannot be exaggerated, and in time increasing numbers of similar modified trows reached London,[6] first via the Thames & Severn, but later also by the Kennet & Avon, for they were built at Wiltshire and Berkshire yards as well as The Bourn.[7]

On the other hand, the carriers were still hampered by the state of the upper Thames and the predominance of eastbound freight. They maintained a weekly service to and from London, but were very reluctant to let boats return empty, so that, as before, goods awaiting shipment piled up on the

One guinea bank note issued by Richard Miller's Brimscombe Port Bank 1818.

wharves at Brimscombe. Their conditions of carriage were far more stringent than the company's; repudiating all liability for loss or damage unless "occasioned by want of ordinary care in the Master or Crew of the Vessel", they limited compensation to ten per cent of the loss sustained.

Of the men concerned, John Baker of Brimscombe Port was one of the few private carriers who had previously used the canal, and he confined his part in the association to the Stourport and Brimscombe trade with which he was familiar.[8] Robert and Lawrence Wyatt of Lechlade and Oxford were men with more than forty years' experience of Thames carrying and appear to have profited from it, for by 1812 they owned nine barges, the third largest fleet on the river west of London.[9] Joseph Brookings of Bablock Hythe, however, was in difficulty in 1811, Sills & Co. refusing acceptance of a draft, and when one of his barges sank at Marlow a month later, Denyer feared he would be ruined; nevertheless, he survived, and appears as owner of one Thames barge the following year.[10]

The fifth, Richard Miller, had gone to Gloucestershire in 1793 as an apprentice in the Brimscombe Port office, but had quitted the company's employ soon after serving his time and hired one of their trows and two Thames barges to carry coal to Lechlade. These he later bought, and within ten years had rented one of the docks at The Bourn and become owner of seven barges, the fourth largest number west of London.[11] Emboldened by success as barge-master and coal merchant, he set up as auctioneer and coach proprietor, running a four-horse coach nightly between Stroud and London, and before the end of 1817 had founded the Brimscombe Port Bank, where, like hundreds of other country bankers at that time, he issued his own notes.[12]

His private pew in Minchinhampton Church, "very excellent & commodious, recently fitted up and lined with green cloth", is redolent of prosperity. But the Stroud and London coaching business was no more secure than water carrying, and to the hazards of these he had added the perils of banking in an age when its principles were little understood and failures frequent. Miller served an area both industrial and agricultural, but at a time when the local clothing industry was losing ground to Yorkshire's and the value of farm produce was falling heavily: in four years the price of wheat and meat fell more than 40 per cent, and in 1822 his Brimscombe Port Bank failed.[13] When Miller's effects were sold, a Thames barge which he had bought for £175, and for which he had refused £150 twelve months earlier, went for £43. It was bought by John George of Brimscombe Port who later became Denyer's son-in-law and the principal trader on the canal, but twenty years later he too was in difficulties and was subsequently declared bankrupt.

One of John George's interests was one-third ownership of a Severn trow in partnership with a gentleman of Bullo Pill and a merchant of Lydney,[14] ports serving the Forest of Dean, which had by then become the main source of the coal shipped along the Thames & Severn Canal. Very soon after formation of the company, the proprietors had shown interest in this coalfield so ideally situated for their trade, but they had found its development inhibited by two factors: restrictive rights and inadequate transport. The Forest was Crown property, and under ancient custom the minerals could only be worked by Free Miners, men born and living within the Hundred of St Briavels who had worked in a pit for a year and a day. Sometimes a 'foreigner' from outside the boundary had invested in a free miner's works or himself had been made an honorary free miner, but both practices were of doubtful legality. So the mines of the Forest, which is very wet underground, had remained small simple workings, entered by levels, drained by adits, but, even so, quickly flooded and soon abandoned because the miners could not afford to install pumps.[15] Nevertheless, in 1787 the Thames & Severn company had made an agreement with Thomas Gilbert, Free Miner, for him to complete at their expense, and make over to them for a consideration of £30, his mine at Tibberton's Folly on Staple Edge Hill (near Soudley), but the agreement does not seem to have had any practical result. As for transport, the Forest is hilly and the roads were few, so the coal was carried on muleback, mostly for local distribution, although small quantities found their way to ports on Severn and Wye. A canal was suggested, but James Black's opinion was that in such country it would show little improvement on the mules. Later on, railways were proposed, but this was opposed by the Crown, and, indeed, there was no real incentive to improve transport until the mines had been developed.

In the early years of the nineteenth century, development came apace.

With the connivance of free miners or in partnership with them, 'foreigners' invested capital. Deep shafts were sunk, pumps installed, steam engines erected, and extensive headings driven.[16] Two railways or tramways were proposed, one from the Severn at Lydney to the collieries at Park End, near the Speech House, and the other from the river at Bullo Pill to the Cinderford mines. But even when the Crown had withdrawn its opposition, parliamentary consent could not be gained, so, wearying of the protracted delay, the promoters of the Bullo Pill line decided to act without it and build a private railway. As one of them came of a family whose members had long exercised great local influence and often been made honorary free miners, and moreover were owners of one of the two estates to be crossed, this was not difficult. The land was leased in the summer of 1807, and the tramway had been "already formed and nearly completed" by 1809 when the promoters successfully applied to Parliament for power to extend further into the Forest and to incorporate the Bullo Pill Railway Company.[17] There is, however, evidence suggesting that part of the line had been in use as early as the autumn of 1807,[18] and certainly it was during the financial year 1807-1808 that the import of Forest coal to the Thames & Severn took the first leap in the phenomenal rise which followed.

At Bullo Pill, there was at first no more than a confined wharf along the riverside with mooring posts for waiting trows and barges,[19] and the Thames & Severn company soon found that coal obtained from Bullo had lain a long time on the wharf and suffered from exposure to the weather; so when the Severn & Wye Railway was opened in 1813 with a sheltered basin at Lydney, they began to draw the bulk of their Forest coal from there instead. In the canal trade, however, Bullo had one supreme advantage: from the Pill to Framilode was no more than 5¾ miles upriver, and barges were able to ply with a minimum of effort, passing up on Severn flood and returning with the ebb. Freights were therefore lower than from any other port whence coal could be obtained, and at Brimscombe in 1811, Bullo coal cost 18s. a ton, Lydney 20s. 6d., Newport 21s., and Staffordshire 23s. 6d.

In the attempt to improve the facilities at Bullo Pill, construction of a wet dock (dated 1818) was begun, but completion languished until the tremendous opportunity for expansion of the Bullo trade in conjunction with the canals was seized by Edward Protheroe, a prominent Bristol citizen, who was emerging as the dominant figure in the railway and mining enterprises of the Forest. Having first invested in the works of his uncle at Park End, Protheroe subscribed heavily to the Severn & Wye Railway, and then, to increase the earnings of the railway — for he maintained that his profits came from the railways, not industry — poured thousands into coalmines, furnaces and forges until he was "by far the largest proprietor" of the local industries.[20] In 1821 he began buying shares in the Thames & Severn Canal, some 400 passing freely

Bullo Pill quay wall, River Severn, in 1952.

Bullo Pill wet dock, built 1818-1827, in 1952.

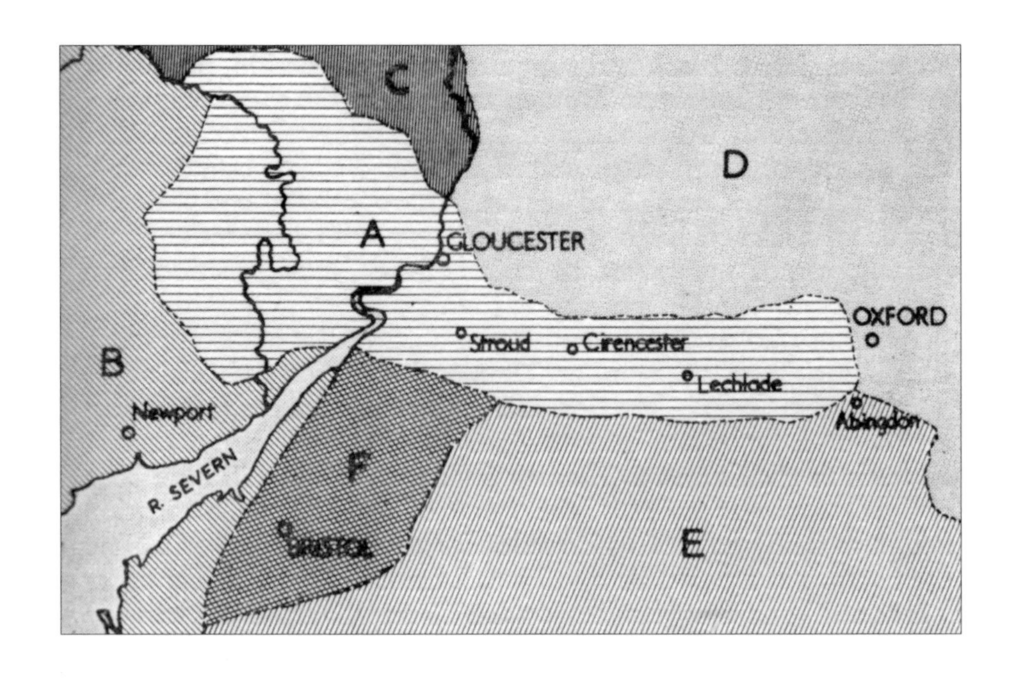

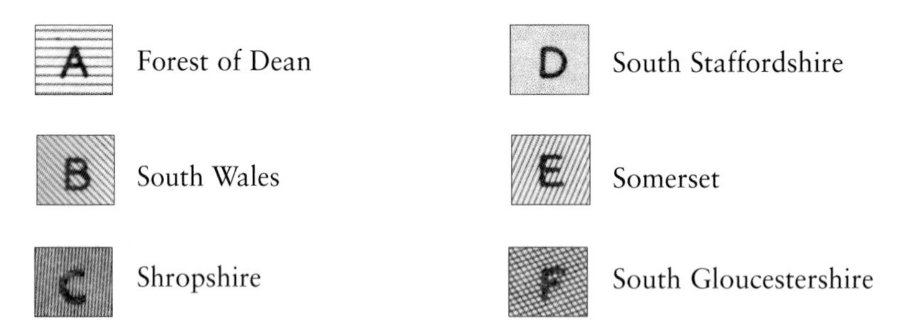

Commercial distribution of coal in the canal area, 1830.

through his hands during the next six years. Four or five years later he acquired the mines and tramway of the Bullo Pill company, and in 1826 formed a new company, the Forest of Dean Railway, to purchase it. He and members of his family held about a third of the shares, but Thames & Severn proprietors were also influential: John Disney held 180 shares and frequently took the chair at the railway company's meetings, and he, John Stevenson Salt and John George were members of the managing committee.[21]

The tramway which became so closely linked with the fortunes of the Thames & Severn was a very simple affair, as, apart from a tunnel 1,100 yards long, earthworks were of the slightest. The single line with passing loops at intervals was laid with plate rails on stone blocks, and meandered along the slopes on a gradient suitable for horses returning with empty wagons from the port. The quays were extended and the dock completed in 1827 at Protheroe's expense, and the high tolls were halved, but the new proprietors remained well content with "the simplicity of the whole concern, and the remarkable economy of the management", although they found that "the steep and crooked track" and narrow tunnel precluded the use of the locomotives they would have liked to introduce in 1838.[22] Indeed, there was cause for satisfaction all round: the exports increased until they were double those of Lydney;[23] the dividends were between 3 and 5½ per cent; the Stroudwater and Thames & Severn canals became the main artery for the distribution of Forest coal; and the average annual import of all coal to the Thames & Severn increased from 17,944 tons in the decade before Forest coal became freely available, to 43,240 in the decade following Protheroe's formation of the Forest of Dean Railway. In two quite exceptional years, 1827-1828, when Forest coal reached a peak not only in volume but also in relation to all other traffic, it amounted to 77 per cent of all coal imported and no less than 60 per cent of the total traffic carried along the canal.

The greater part of this coal was, of course, carried to the company's own wharves, but by means of drawbacks the market was pushed down the Thames into the area supplied by the Oxford Canal, moderate quantities being sold in Oxford and Abingdon, small amounts as far down as Reading and even Henley. The area within which the sale of Forest coal predominated was, however, a belt some ten or eleven miles broad astride canal and river and ending just short of Witney, Abingdon and Oxford.[24] Within this area there was little competition except in Cirencester where Sodbury coal continued to find a market.

In course of time, the Forest trade led to the development of yet another distinctive type of vessel, the Stroud barge. Designed to carry 60 tons of coal to Stroud, Brimscombe and Chalford, it was flat-bottomed and double-ended, 68 feet long, 12½ feet wide[25] — like the hybrid trows, short enough for the locks below Brimscombe Port and narrow enough for those above. Little is

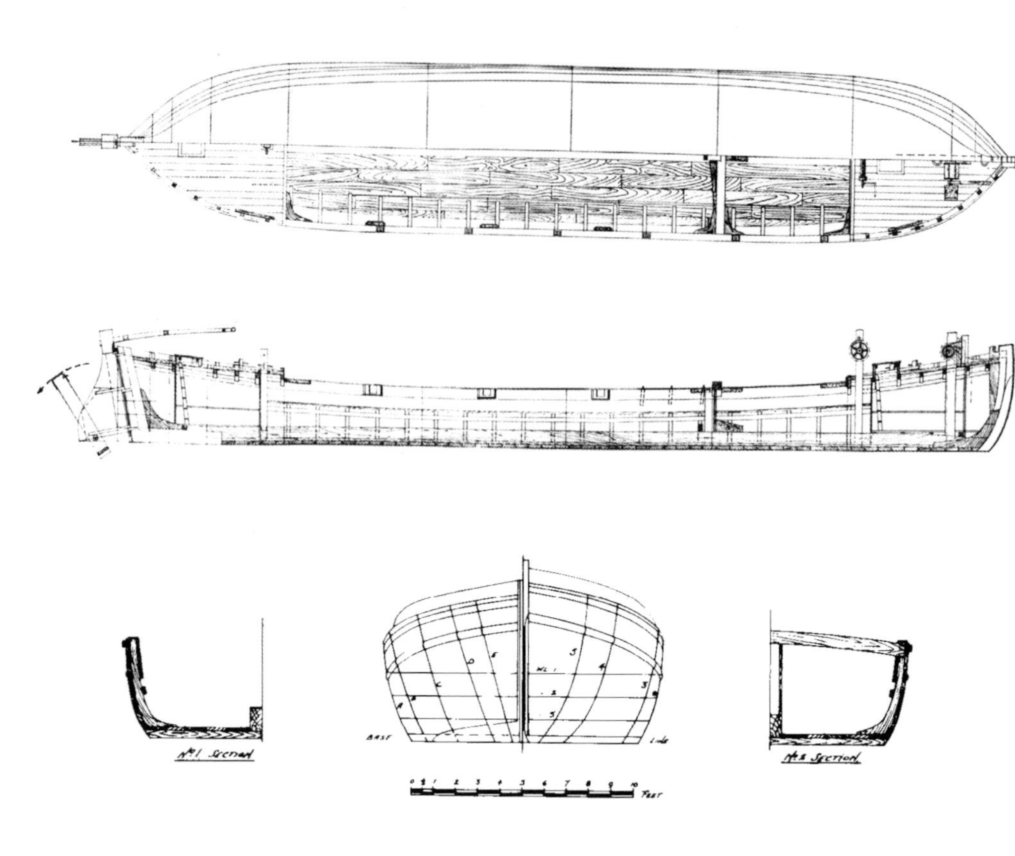

Dimensions: L O A 68' 4" x 12' 6" Hold Depth 3' 11".

N.B.

Barges were towing barges, no sails being used, but it is possible that a square sail was used when the barges were trading beyond the canal limits. On the River Severn, at Framilode, or adjacent canal waters.

Plan and section of the Stroud barge *Perseverance* owned by James & Sons, Chalford.

known of their origin, but it seems certain that originally they had no masts and were therefore intended solely for the Bullo trade in which they were moved on the Severn by the tides. In later years, however, some were adapted to carry sails and work to other ports.

Concurrently the fortunes of the Thames & Severn company were further improved by forming a junction with the Wilts & Berks Canal, so at last ending their entire dependence on that deplorable section of the Thames navigation above Abingdon. As early as 1797, when the Wilts & Berks was in the first stages of construction, James Black had suggested such a junction, proposing a line from Inglesham to Shrivenham, where the routes of the two were only 6½ miles apart. The idea was well received until the Bishop of Durham, who had already once worsted the Wilts & Berks, protested that his Beckett estate would thereby become 'an island' between two canals, and Edward Loveden disclosed his opposition to any such connection, supporting instead a proposed extension of the Grand Junction from Aylesbury to the Thames at Pinkhill.

Although Loveden argued that the Wilts & Berks was a narrow canal, unable to take either Thames barges or hybrid trows, whereas the Grand Junction was a broad one, he was in fact moved entirely by self-interest. Had that proposal been adopted, all trade from the Thames & Severn would still have had to use twenty-five miles of the worst of the Thames navigation and in so doing pay toll to Loveden for passing Buscot Lock. Now, the charges at Buscot were unique, for not only was 1s. for every 5 tons as high a charge as any on the Thames, but Buscot was one of the very few locks where toll was exacted on passage up the river and again on return[26] — and even when the canal trade had been most in need of encouragement, he had refused to abate his demands one jot. In time, Loveden's opposition to the junction with the Wilts & Berks became implacable and his methods utterly unscrupulous.

Negotiations with the Wilts & Berks were reopened in 1810 when that canal and the Kennet & Avon (which it joined at Semington) were both nearing completion. In spite of acute financial embarrassment,[27] the Wilts & Berks had announced grandiose plans for extension eastward to the Grand Junction, westward through the Kingswood coalfield to Bristol, and northward by the Severn Junction Canal to Cirencester. It was an audacious set of proposals, threatening the Thames & Severn with competition in the coal trade of the countryside, but offering them complete independence of the River Thames in return for the sacrifice of half their mileage. It is surprising, in view of the impecunious state of the parent company, that Disney and his colleagues took the proposals seriously, but they stood their ground, and at a joint meeting of the two managing committees in July, the Wilts & Berks agreed to drop the Bristol extension, to adopt for the Severn junction a line from Latton to Swindon, and to encourage the suggested link with the Grand Junction, mere talk of which might, it was thought, stimulate the Thames Commissioners to action.

This it quickly did, but not to the action wished for. Concurrently there were two other schemes intended to divert trade from the Thames: one for the Kennet & Avon and Basingstoke canals to connect, and the other for revival of the proposed London Canal from Reading to Isleworth. Before the year was out, two reports were printed, attacking the canal projects and remarking complacently on "the present sufficient and still improving State" of the river navigation. The author of the first was Loveden whose lucrative little lock was threatened by one of the projects, and of the second, Colonel Page of Newbury, sole proprietor of the Kennet Navigation, whose very profitable monopoly was endangered by the other. Alleging that the Wilts & Berks and Grand Junction canals, both of which were notoriously short of water, intended to draw supplies from the river via the Severn Junction, Loveden concluded with a sentimental appeal to resist "the conspiracy against old Father Thames".[28] It was an appeal astutely calculated to stir the emotions of a far wider public than his fellow commissioners to whom it was addressed. When the Bill for the Severn Junction was introduced in March 1811, a flood of petitions quickly showed the strength of the opposition, and as the Thames & Severn had meanwhile discovered that their own powers would have to be amended before the junction could become effective, proceedings were dropped for that session.[29]

Loveden saw a last opportunity of protecting his interests. Buying "all the shares offered" for public sale (ten in May 1811) and taking advantage of the customary meagre attendance at the proprietors' assemblies, he packed the next meeting (in January 1812) with his nominees, asserted that the Thames Commissioners were about to introduce a Bill for improvement of the river above Abingdon, and successfully moved a resolution opposing the Latton junction. John Disney and two other members of the committee were present but abstained from what they realised would be an ineffectual casting of their votes. Loveden's triumph was short lived: he had scotched the measure for that session, but no more. It soon appeared that the Thames Commissioners were seeking no such power as he claimed, and if the canal proprietors needed further proof that Disney's policy was right, it was provided by the news that the Wilts & Berks intended to drop the Severn Junction Canal and promote their Bristol extension. Losing no time, Disney sent a circular letter to the proprietors,[30] explaining the surprising result of the meeting held six weeks earlier, telling them of the subsequent developments, and summoning them to meet again in a month's time.

Disillusioned but not yet defeated, Loveden then sought to discredit his opponents by accusing the Wilts & Berks of bad faith and Disney of having "taken a retainer from that company". However, when the canal proprietors met again in March, they approved their chairman's policy, and at a meeting held later in the year to approve details Loveden made his last appearance. Recognising defeat but unable to conceal his chagrin or curb his anger, he defaced the minute book, the company's clerk recording in the margin: "After this Meeting was over Mr Loveden came to the Table, took the Book out of my hands, and struck his name out, saying he would not have his name appear when he did not approve the Resolutions".

Although the Bill for the North Wilts Canal, as the proposed junction had been renamed, carefully protected Old Father Thames by forbidding the company to take any water from the river or from the Thames & Severn Canal except during construction, it encountered vigorous opposition. Nevertheless both it and the Thames & Severn company's amending Bill received Royal Assent on 2 July 1813.[31] The promoters named in the Act were mostly proprietors of the Wilts & Berks, but amongst others were the Bullo Pill Railway Company and half-a-dozen proprietors of the Thames & Severn. But the public were so reluctant to take up shares that the two parent companies were obliged to invest in their corporate capacity, the Thames & Severn putting up £5,000 (the whole of their sinking fund at that date) and in return being entitled to appoint a member of the North Wilts committee. Reckoning that their investment would be well rewarded by an increase in their own receipts from tolls, they undertook to encourage the flow of traffic by granting a drawback of 2s. 6d. a ton on coal passing via the junction, and to accept far lower minimum charges than before, outward bound boats paying toll on at least twenty tons and incoming ones on their actual lading.[32]

The North Wilts Canal, 8 3/8 miles long, was surveyed by William Whitworth, Robert Whitworth's son.[33] Construction of the junction basin at Weymoor Bridge and of the aqueduct across the Thames at Latton, both of which were the property of the Thames & Severn, was supervised with characteristic thoroughness by John Denyer, who prevented stoppage of Latton Mill and a claim for damages by temporarily diverting the millstream to flow through the canal, and later, when bad weather flooded the excavations, set horses to turn an Archimedian screw day and night so that masons could continue work on the foundations. Double stop-gates were provided, as specified in the North Wilts Act, one pair being under the control of the Thames & Severn, who were authorised to bar passage if the water level in the junction canal should fall below their own. After the North Wilts had been opened on 2 April 1819 and merged in the Wilts & Berks two years later,[34] this arrangement led to a great deal of recrimination, shortage of water being a matter on which the managements of both companies were extremely sensitive. On several occasions when the Thames & Severn exercised their right, the Wilts & Berks retaliated by forcing the gates to pass boats from their line, and by venially, but illegally, barring passage when their water level was the higher.

Through the junction, narrow boats of the midlands type carrying 20 to 30 tons began to use the Thames & Severn. At first glance it appears impossible that vessels 70-72 feet long could pass the 68-foot locks below The Bourn, but there is nothing in the minutes to suggest that lengthening was ever thought necessary, much less that it was done to seven locks, interrupting navigation for a long period which could not have failed to find mention. As it is the inward opening of the lower gates which limits the length of a vessel of full width, the length of a lock is measured from the upper sill to the recess in the lock wall filled by those gates when open. But these locks were wide and the boats narrow, so that by manoeuvring the vessel sideways in the lock, it was possible to open — or close — first one lower gate and then the other.

Inevitably, narrow boats began to oust the traditional river craft for which the canal had been designed. By 1839, of 1,843 boats that passed over the summit level, some six a day, 1,526 were narrow canal boats, 317 were trows, but none were Thames barges. That the company benefited by accepting boats with half the capacity of those previously used, and by sacrificing a large proportion of the earnings of nine miles of canal, is a measure of the handicap previously imposed on them by the navigation of the Thames above Abingdon. Complete independence of the Thames by connection with the Grand Junction was never achieved, although it was again discussed in 1817-1819 and 1828.

After the Kennet & Avon and Wilts & Berks opened towards the end of 1810, the considerable Bristol and London trade of the Thames & Severn shrank to vanishing point, and, despite all efforts, never recovered. Receipts fell by nearly 38 per cent in the four ensuing years, and the dividends, which had been 30s. on both classes of share for 1810 and 1811, dropped to 10s. on the Black shares in 1812 and to 17s. on the Red in 1816. That year Red shares sold for no more than £10. However, the phenomenal rise of the Forest coal trade and the building of the North Wilts Canal led to recovery, and when three Red shares and ten Black were sold in 1819, the former realised £35 10s. each and the latter, £17 10s or £18.

Meantime the company's western outlet, the Stroudwater Navigation, approached the jubilee of its opening and the centenary of its first Act, and in the hands of an unenterprising management retained two early eighteenth century features which were outmoded and irritating — the towing-path unusable by horses, and the exorbitant toll of 3s. 6d. to Stroud, 5¼d. per ton per mile. Although only 37 per cent of the coal brought up the canal paid the full tonnage, the remainder passing at reduced rates to the Thames & Severn, the company were so prosperous that in 1819-1824 dividends averaged £24 5s. The management would therefore have been wise to listen to complaints about the cost of coal in Stroud, a matter of special concern to the growing number of mill owners using steam to supplement water power;[35] but as they showed no sign of doing so, a group of manufacturers decided to force the issue.

It was not difficult to find a way: the Stockton & Darlington Railway was being built, the Liverpool & Manchester had been projected, and it was obvious that discussion of a Stroud and Severn railway, if well enough supported, would

Narrow boats on the Thames at Wallingford in 1810 when the bridge was being repaired. Note the Archimedian screw drawing water from the coffer-dam.

force the Stroudwater proprietors to negotiate. At a meeting in September 1824, those present decided to support such a railway and to appoint a committee, but to take no further steps pending the outcome of negotiations.[36] Shaken out of their complacency at last, the navigation proprietors also appointed a committee, empowered to 'conclude a treaty'. This proved more difficult than supposed, for the railway committee insisted that the canal toll should be reduced to 2½d. per ton per mile, which the navigation committee considered "so unreasonable that they [could] not possibly accede to it". They did, however, reduce it to 2s. 6d. (3¾d. per ton per mile), whereupon some landowners announced that a further cut of 6d. would tip their influence against the railway scheme. Nevertheless, although the canal proprietors agreed to this, the railway party decided to present a Bill in Parliament.

Up to this point, the Thames & Severn had noted the proceedings without sympathy for the Stroudwater's plight, but also without alarm. The promotion of a railway intended to serve not only Stroud, but also Brimscombe Port, was a different matter, and, thoroughly aroused, they joined the Stroudwater in active opposition. Petitioning against the Bill, the Stroudwater company claimed with justice that their dividends ought to be considered in relation to the sum they had actually raised rather than the nominal capital, and added the sentimental appeal that the railway would "ruin many Widows and Orphans, who are Proprietors of Canal Shares". The Bill did not get far, second reading being 'postponed' by a large majority in April 1825.

The matter of the towing-path was raised later that same year by John Disney, who pointed out that when the Gloucester & Berkeley should be finished, the navigation below Brimscombe would be the only part unusable by horses in a long line of communication between the north, midlands, and east. The Stroudwater conceded that improvement was 'expedient', but having seen their revenue cut and dividend halved, they were in no hurry and agreed only "to adapt the present Towing path to the passage of Horses . . . as a Temporary experiment", and in this niggardly spirit, the work was done, though not until the summer of 1827 and after the eagerly awaited event which had prompted it.[37]

The Gloucester & Berkeley, a ship canal 16½ miles long avoiding some thirty miles of dangerous navigation in the Severn estuary, was begun in 1794, but proved such a difficult and costly undertaking that it long remained unfinished. Few, if any, parts were more complex than the crossing of the Stroudwater Navigation, for a deviation from the original plan had shifted the point of intersection westwards from above the Stroudwater's Whitminster Lock, where the two canals would have been on the same level, to Saul where the level of the older canal was lower than that of the newer. A temporary junction was formed on the existing level in February 1820, opening communication between Gloucester, Stroud and the Severn at Framilode, but the permanent works involved altering Whitminster Lock, forming a new bed for the Stroudwater 5 feet 10 inches above the former level, building a new lock on the western side of the junction for the exit of the Stroudwater on its old level, and erecting an elaborate system of stop gates to prevent loss of water in any direction from the three level arms. The work was done in the summer of 1826, compensation being paid to the Stroudwater and Thames & Severn companies while navigation was interrupted, but as parts of the works twice failed and had to be renewed, the operation took ten weeks, much longer than expected, and with rising impatience at the protracted delay, John Denyer paid repeated visits, once persuading the Stroudwater committee to accompany him, and each time urging those concerned to be as quick as possible.[38]

When at last the Gloucester & Berkeley Canal was opened on 26 April 1827,[39] the Thames & Severn company expected great advantages to follow, particularly the shortening of journey times. There was need: the maximum economic speed on a canal was reckoned to be 2½ mph; and overall speed was very much lower when canals were open only by day and river navigations were involved. The railways then being ardently promoted could do very much better than that. How could the times be reduced? Following experiments with steam boats around the turn of the century, development of these seemed likely to offer a way. On the Forth & Clyde Canal, William Symington's paddle steamer *Charlotte Dundas* had managed 2½ mph towing three 60-70 ton vessels in 1801, and the Duke of Bridgewater, who even earlier had tried steam towing, had been so impressed by the possibilities that he ordered some half-dozen steam tugs similar to Symington's shortly before he died. Unfortunately his trustees, probably sharing the view that the damage inflicted by churning paddle wheels on the banks and bed of a canal outweighed any possible advantages, cancelled the order, and the impetus the Duke's interest would have given was lost.[40] Not all canal managements shared the general prejudice, certainly not that of the Thames & Severn, for John Denyer, commenting on a Mr McCurdy's unsuccessful attempt to reach Oxford in a steam boat with a 4 hp engine in 1827, remarked he was sorry it "had been so unfortunate as it may tend to deter others".[41]

Shortly after, there was a proposal to use steam on the Thames & Severn itself. This followed David Gordon's voyage in a small steam boat from Paddington to Manchester in 1828, when he is said to have proved "the possibility of employing steam-vessels without injury to the canals".[42] The Following year, Gordon called at Salt's house in Lombard Street while the proprietors of the Thames & Severn were in session, and distributed prospectuses of a Thames & Severn Steam Navigation Company to operate from London to Gloucester and Worcester via Stroud, as well as to Bristol via the Kennet & Avon, using on the rivers a steamer carrying passengers and light goods and towing a pair of narrow boats, and on the canals, a steam tug narrow enough to enter the locks with a single narrow boat.[43] Unfortunately that led nowhere.

About the same time the distinguished engineer William Fairbairn (later Sir William) took up the question of speed on behalf of the Forth & Clyde Canal. Impressed by the remarkable experiment on the Glasgow Paisley & Ardrossan Canal when a light passenger boat drawn by one horse skimmed over the surface at 12 mph, he believed that steam could do just as well, and about 1831 he built for the Forth & Clyde a tug with a paddle wheel mounted amidships, but the disappointing results convinced him that it was only on the broader canals (and the Thames & Severn was one of those he named) that there was any hope of increasing speed, and then only to some 3½-4 mph.[44]

Little further attempt was made to use steam on canals until many years later.[45] Having ceased to act as carriers, canal proprietors had little incentive to experiment themselves and less to revet banks so that others might, while very few of the private carriers could have made sufficiently intensive use of a tug even if they had the means to acquire one. It is significant that none was introduced on the Thames & Severn until the company again became carriers in 1875.

A simpler and more economic way of shortening journey times had been developed on the Grand Junction by using 'light-boats' or 'fly-boats' carrying small consignments of valuables and general merchandise, which, travelling night and day, using relays of horses and a relief crew, were by 1819 maintaining a regular and reliable service covering 101½ miles at an overall average speed of just over 2 mph.[46] Once the North Wilts and Gloucester & Berkeley canals had been opened and there was a horse towing path below Brimscombe, it was possible to introduce fly-boats between London and Gloucester via the Thames & Severn. Started on 1 November 1829, this was an immediate success: in the first year 258 boats passed, and in time three firms, all established Witney carriers, took part — Parker & Foster, George Franklin, and Bowerman, Son & Mason — advertising between them services starting on four days a week in each direction. The fastest left Gloucester on Monday morning and, helped by the flow of the Thames, completed the journey in four days, but all the others included a Sunday when the boats lay idle, and took five to seven days to reach London, and six or seven to return. By arrangement with other carriers, goods were accepted for carriage to a wide range of destinations in Gloucestershire, the neighbouring counties, the upper Severn Valley, and Wales.

Just as the canal company had helped the associated carriers of twenty-five years earlier by charging them a consolidated toll of 4s. 6d. a ton, so they, did in 1830, but at the still lower rate of 2s. 6d. a ton on cargoes carried to or from Pangbourne Lock or places lower down the Thames. Of the three operators, only Franklin appears to have advertised his rates, but these, which included delivery to any part of London, were very reasonable in comparison with those charged by the company's own services some thirty years before: from London, wool, hops and tea paid the highest rate, 60s. a ton to Brimscombe and Stroud, 55s. to Gloucester, but a large assortment of commodities paid the lowest rate of 35s. a ton; in the reverse direction, a greater variety of rates applied, from 16s. to 55s., but as with the westbound charges, some were lower from Gloucester than Stroud, showing that the fly-boat carriers had to compete with coasters.

These carriers showed far greater readiness to accept risk and responsibility than their predecessors had done, and as the conditions advertised by all three show only minor differences and at least one of them operated over a number of other canals, it seems probable they conformed to some general standard. They promised "particular care . . . of valuable and hazardous Goods", but would not be held accountable for them, nor for parcels exceeding £5 in value (Parker & Foster's limit was £10), unless the contents were declared "and paid for accordingly". "Brittle and Combustible Goods" were carried only "at the entire risk of the Owners", and Parker & Foster added rather naïvely that they would be responsible "neither for leakage of Casks, nor live animals". Furthermore, "should navigation be impeded by frost or high floods" all three carriers were prepared, if required, to arrange conveyance by land at moderate rates.

Stimulated by speed and regularity, the flow of trade between Gloucester and London grew greatly. In the twelve months before the introduction of the new service, only 502 tons passed, but in their first year the fly-boats carried 1,523 tons, and by mid-1834 about double this amount. But the average lading of the fly-boats was very low, only 7½ tons as against 26 tons of other vessels passing eastward off the canal, and the earnings were not enough to satisfy either the canal company, who doubted whether the receipts sufficed to offset the increased wear and tear of locks and consumption of water, or to sustain, three competing carriers. All three were presently in trouble; in the middle of 1833, Denyer had difficulty in getting settlement of their accounts, and cheques presented by Parker and Franklin were dishonoured. Nor was this a temporary embarrassment, as the following year Parker remarked that from his carrying over some ten or twelve canals he had "got and lost a large Fortune". There was only one solution, and in 1834 Bowerman took over Franklin and was then himself absorbed by Parker, whose business was merged in an inland carrying company three years later.

Although the public may not have benefited from the amalgamations — Denyer expected charges would be increased — traffic continued to grow and operating results certainly improved: in the year 1839, 372 fly-boats carried 5,162 tons between the two cities, the average load had increased to 13.9 tons, and the canal company's receipts had risen from 18s. 3½d. to about 34s. per boat. However, there had again been difficulty in getting payment of the account, and in 1840 it was learnt that the carrying company had ceased to

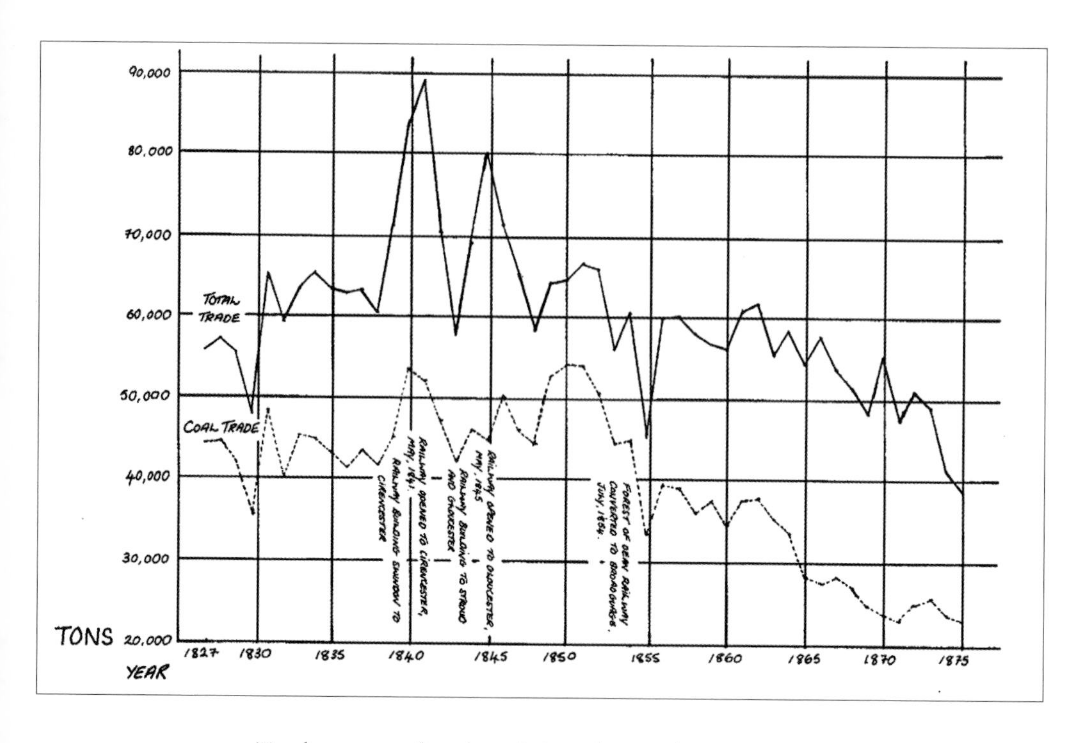

Total tonnage of coal carried on the canal, 1827-1875.

exist and that the business was passing into yet other hands.

As their eastbound and westbound cargoes differed by little more than 5 per cent, the fly-boats had realised one of the Thames & Severn company's hopes, the creation of a balanced trade with London. Yet it was a very small proportion of the whole, only about 6½ per cent. Over the years 1827-39, 91.7 per cent of the total trade of the canal moved eastward, 70.8 per cent was coal (46.6 per cent from the Forest of Dean); and of this mass of eastbound freight, about 43 per cent never left the Golden Valley, 25 per cent was to the company's other wharves, a little more than 16 per cent passed to the Wilts & Berks at Latton, and a little less than 16 per cent to the Thames at Inglesham.[47]

The Thames & Severn Canal, far from developing into the artery of east-west trade expected by its projectors, had become principally a distributor of coal from the Forest of Dean to the homes and factories of the Stroudwater valley.

REFERENCES

1. Registrations of the Port of Gloucester 21 March 1804.
2. Farr, G., *Chepstow Ships*, 1954, pp. 75-6, Ketch *Newport* built at Brimscombe 1799, registered at Chepstow 27 August 1806.
3. See Baddeley, W. St.C., *A History of Cirencester*, 1924, pp. 317, 321.
4. Allnutt, Z., *A New List of Barges on the Thames Navigation*, 1812.
5. GCL, Printed letter "Thames & Severn Canal, 1805".
6. Allnutt, Z., *Useful & Correct Accounts of the Navigation, of the Rivers and Canals West of London*, p. 13, "The Severn Trows also frequently go throughout to London" — the work is undated, but was written after completion of the Grand Junction Canal in 1805 and before that of the Kennet & Avon in 1810. See also Storer & Brewer op cit, 1824, p. 202. Hall, *The Book of the Thames*, 1859, p. 112 note. Thacker op cit vol. 2, pp. 350, 466.
7. Farr, G., in *The Mariners' Mirror* vol. 32, pp. 88-9.
8. Baker appears as owner of two Severn craft in 1795-7, but neither then nor later as owning any Thames barges — see Register of Barges, trows, etc., 1795-7, GCRO, and Allnutt, *A New List of Barges*, 1812.
9. Commissioner, op cit, 1767, pp. 41-2. Allnutt, *A New List of Barges*.
10. Allnutt, *A New List of Barges*.
11. Ibid.
12. An original guinea note dated 1 December 1817 is in GCRO, and another of 1818 is in the possession of the National Westminster Bank, Stroud.
13. I am indebted to T. Hannam-Clark for various notes on Miller, and see Fisher, P. H., *Notes & Recollections of Stroud*, 1871, pp. 109-112,

Bankers' Almanack, Social England vol. 6, p. 112 for the fall in prices.

14. Farr, G., *Chepstow Ships*, p. 117.
15. For conditions in the Forest of Dean, see Rudder, S., *A New History of Gloucestershire*, 1779, p. 36; PP 1830 *Accounts & Papers* vol. 29, p. 197; PP 1835 Reports vol. 36, pp. 122, 156, 165, 173, 174; Nicholls, H. G., *The Forest of Dean* 1858: Nicholls, H.G., *Iron Making in the Olden Times*, 1866, pp. 67-8; Galloway, R. L., *Annals of Coal Mining* vol. 2, pp. 146-8; Kerr, R. J., *The Customs of the Forest of Dean* in *Transactions of the Bristol & Gloucestershire Archaeological Society* vol. 43, p. 73.
16. For the activities of the 'foreigners', see PP 1835 Reports vol. 36, pp. 122, 157-160, 163-5, 173-4, 180.
17. Act 49 Geo. III cap 158. The family was that of Roynon Jones see Nicholls, H. G., *Personalities of the Forest of Dean*, 1863, pp. 76-81, and Nicholls, H. G., *The Forest of Dean* 1858, pp. 66, 71.
18. GCRO Q/RUM 22, a plan deposited in September 1807 refers to the "Present Railway".
19. GCRO Q/RUM 73, deposited plan dated 1821 shows the extent of the early accommodation.
20. PP 1835 Reports vol. 36, pp. 159, 172, 175. Nicholls, *The Forest of Dean*, p. 107. Beavan, A. B., *Bristol Lists: Municipal and Miscellaneous*, 1899, p. 306.
21. Act 7 Geo. IV cap 47. Forest of Dean Railway General Meeting Book 1826-1852.
22. Forest of Dean Railway General Meeting Book, 30 August 1826, 17 January 1838, and FDR Committee Book 13 April, 11 October, 1827.
23. See figures quoted in PP 1871 Reports vol. 18, p. 944.
24. PP 1830 Reports vol. 8, following p. 397 is "A Map showing the Geological Position and Commercial Distribution of the Coal of England and Wales".
25. Drawings of the Stroud Barge *Perseverance* were made in 1936 and are now in the Science Museum, South Kensington.
26. For Buscot Lock, see Vanderstegen op cit, pp. 5, 42, and GCRO Sills's Report of 8 December 1796, List of Tolls.
27. See, for example, Hadfield, *Canals of Southern England*, p. 157.
28. *Two Reports of the Commissioners of the Thames Navigation, on the Objects and Consequences of the several projected Canals, which interfere with the interests of that River: and on the present sufficient and still improving State of its Navigation*, 1811. See also *An Authentic description of the Kennet & Avon Canal* 1811, p. 24, and Hadfield, *British Canals*, pp. 165-6.
29. JHC vol. 66, many entries between pp. 67 and 230. Broadsheet, Proposed *Severn Junction Canal* 1811 in GCL.

30. GCL letter from J. Disney to the Proprietors of the Thames & Severn Canal 18 February 1812.
31. JHC vol. 68, many entries between pp. 91 and 631.
32. Acts 53 Geo. III cap 181 & 182. Priestley op cit, p. 718.
33. See report appended to broadsheet in GCL, *Proposed Severn Junction Canal* 1811, in which W. Whitworth mentions having surveyed the water supplies of the Wilts & Berks "in conjunction with my late father".
34. *Report to the Allied Navigations Re Thames & Severn Canal* 1894, p. 25. Priestley, op cit, pp. 718-9.
35. See, for example, Playne, A. T., *A History of the Parishes of Minchinhampton and Avening*, 1915, p. 153.
36. For the whole story of the railway proposals, see the interesting material in GCL (JF.16.106 1-10 and JF.14.83) as well as SWN committee 20 & 28 October & 2 November 1824, 11 May 1825.
37. SWN committee 5 July & 18 October 1825, 27 March & 25 April 1827.
38. Phillips op cit, pp. 302-3, 403-5. Gloucester & Berkeley Canal minute book 4 & 11 January & 24 March 1820, 13 November 1826. SWN committee 27 December 1819, 4-5 January, 7 March 1820, 15 August 1826. GCRO "A Plan of the Junction of the Stroud, Gloucester and Berkeley Canals. Copied from a Survey made by Mr H. Sutherland, 1824. Drawn by W. Clegram 1826". A letter from J. Denyer to J. S. Salt of 22 May 1826 (GCRO) mentioned that the Stroudwater would be stopped for at least six weeks "for the purpose of rebuilding a Lock & completing the junction with the Gloucester & Berkeley Canal". The canal was actually closed from 29 May to 5 August.
39. GJ 28 April 1827.
40. Ellesmere, Earl of, *Essays on History, Biography, Geography, Engineering &c.*, 1858, No. 7 *Aqueducts and Canals*; Fairbairn, William, *Remarks on Canal Navigation, illustrative of the advantages of the use of steam . . . .*, 1831, pp. 5-6. Rees, *Cyclopaedia*; Canal. Smiles op cit vol. 1, pp. 409-410, 448-9, vol. 4, pp. 448-9.
41. Letter from J. Denyer to J. S. Salt 19 February 1827 in GCRO.
42. *Gentleman's Magazine* vol. 98, p. 268.
43. A copy of the prospectus in the William Salt Library, Stafford, is endorsed in J. S. Salt's hand: "These papers were circulated by David Gordon, who called in Lombard Street on 28th April 1829 the day of the General Assembly".
44. Fairbairn op cit 1831, particularly pp. 6-28, 30-33, 49-51, 81, 83, 90-1; Pole, William, *The Life of Sir William Fairbairn* 1877, pp. 133-145.
45. For some idea of what was done see Dupin op cit vo1. 2, p. 211 note, and Hadfield, *British Canals*, p. 53.
46. Hassell op cit, 1819, p. 88. Rees, *Cyclopaedia*, Canal refers to 'light-boats'

passing by night in 1805.

47. Summaries of the fly-boat trade are given in a letter from J. Denyer to William Herrick 28 February 1831, and in a "Synopsis of the London and Gloucester Fly Boat Trade on the Thames & Severn Canal from the 1st Nov 1829 to 28 June 1834" in GCRO. Other figures have been drawn from the Fact Book, and the Siddington Return sent weekly by the clerk-of-the-works to the treasurer which recorded, inter alia, the number and type of vessels which passed, the nature, weight and destination of the cargoes. I have extracted the figures for the year 1839, one in which the trade of the canal approached its peak, and the last before the returns were seriously affected by material carried for the building of the railway between Swindon and Cirencester.

CHAPTER ELEVEN

ZENITH OF EFFICIENCY

The nineteenth century annals of the Thames & Severn Canal record many instances of long and faithful service of officers and servants, instances which are not only remarkable in themselves but of great significance in the company's history.

John Disney, the first permanent chairman whose resolute pursuit of three aims reform of the finances, development of the Forest coal trade, and junction with the Wilts & Berks — had improved the company's prospects beyond all probability, resigned in 1828. Recording their appreciation of his "indefatigable attention . . . to their affairs" for twenty-three years, the proprietors presented him with an inscribed piece of plate worth £200. He continued to serve as a member of the managing committee, and maintained his keen interest in the Forest of Dean Railway, but his chief interest thereafter was the collection of classical antiquities. He is said to have shown "more zeal than discernment", but he was nevertheless an archaeologist of distinction, Fellow of the Royal Society, and founder of the Disney Chair of Archaeology at Cambridge.[1] His successor, William Parry Richards, was a man of less forceful character, but his policy of restraining dividends for improvement of the property and to build up a contingency fund was no less sound and beneficial. Unfortunately when he retired, the proprietors were in no position to give him anything but their thanks.

After nearly twenty-nine years the proprietors at last appointed a clerk, John Lane, son of Thomas Lane, a promoter of the Thames & Severn who had often acted as their legal adviser. Like his father, John Lane was a solicitor and clerk of the Goldsmiths' Company.[2] Appointed in April 1812 at a salary of £50 a year, he was never a resident full-time officer as was the Stroudwater's clerk, but for thirty-eight years managed the company's business from his London office.

The office of clerk-of-the-works, so ably filled by Samuel Smith while the canal was being built, was held for sixteen years by John Proctor, who had first worked for the company as a day labourer in 1785. In 1814 he was succeeded by another and quite different Charles Jones, who held the position for more than half-a-century.

In fifty-five years, then, there were only two chairmen, and for thirty of those, there were the same treasurer, clerk and manager, and only two clerks-of-the-works. Moreover, the five principal officers, Richards, Salt, Lane, Denyer and Jones, worked in unbroken partnership for 14½ years and averaged forty years' service apiece — a very remarkable record. Other employees also served as long. Thomas Toward, having unsuccessfully asked for a pension in 1824, stayed on at Thames Head until the beginning of 1833, completing more than forty years as engineman on a salary which had risen meanwhile from fifty to eighty guineas. Three lock-keepers, Pash, Ridley and Meacham, served 25, 40 and 50 years respectively, William Dicks, a clerk in the Brimscombe office, more than fifty years.

The staff was small, the incidence of long service high. The records suggest what one would infer, that employers and employees were personally known to one another and bound by a common loyalty. Many entries in the minute books show that the proprietors rewarded good work and readily alleviated the distress or succoured the old age of those who served them faithfully. John Proctor, when infirm and no longer able to act in the dual capacity of clerk-of-the-works and toll collector at Siddington, was relieved of his works duties but paid the same salary. Benjamin Haines, the carpenter, "by age and infirmity reduced to a helpless state", was granted 9s. a week for life; and when the three faithful lock-keepers became too old for their work, two still capable of lighter duties were paid 6s. a week, and the third a pension of 5s. "The faithfull service of the Clerks in the Office, one of whom had been in the Company's employment nearly 40 years, the other about 8 years", were once rewarded, on Denyer's recommendation, with bonuses of £7 and £3 respectively, and ten years later the senior of the two, by then aged, ill and helpless, was granted an annuity of £40. Toward's successor, after no more than 2½ years, was given £10 on account of his "useful services and good conduct . . ., as well as his misfortune in the loss of his Cow". Jones was given £15 after a severe illness in 1835, a gratuity of £30 in 1839, another of £50 in 1841, a rise in salary from £150 to £200 a year and a share of profits in 1842, and when the time at last came for retirement, a pension of £90 a year. Denyer, whose salary rose by stages from £230 to £300 a year, also received a gratuity of £30 in 1839 and a share of profits three years later, and when he died suddenly, leaving little for his widow, the committee suggested paying her a small annuity of £25 which the proprietors promptly doubled.

As a contemporary remarked that the Duke of Bridgewater's provision for employees "whom age or accident . . . had rendered unfit for service" was "most meritorious",[3] it is probable that canal proprietors were seldom as generous as those of the Thames & Severn and if the rewards now seem small, it must be borne in mind, not only that the value of money was then far greater, but that the wants of mankind were very much simpler.

It is quite evident that the loyalty and benevolence permeating proprietors and staff were due principally to the character and influence of John

Stevenson Salt. His concern for the canals with which he was associated far transcended that arising from financial commitment alone. It sprang from something much more personal, perhaps a feeling for canals somewhat akin to that proud attachment to railways which developed in later generations of railwaymen from directors downwards. Only some such feeling could account for Salt's aversion to the railways which increasingly threatened the canals, an aversion of such intensity that never in his life would he travel by train on the frequent journeys between his London residence and his country home near Stafford. Moreover, every week John Denyer sent his cash account, and the clerk-of-the-works his return of conditions on the summit, direct to the treasurer in Lombard Street, and if Salt had not taken a close personal interest in the canal, its employees, and the locality, Denyer would not have written his informative covering letters, the clerk-of-the works would not have scribbled long notes on his return, nor — and this is indeed remarkable — would Salt have preserved these documents, with the result that at some later date the *originals* were sent to take their place amongst the archives at Brimscombe.

Some of Denyer's letters mention no more than the amount of the remittance. Most describe the state of the works, the effects of the weather, the condition of the coal trade. Sometimes, underlining freely, he gave local news, commenting on distress in the clothing industry, strikes among the weavers, enclosure riots in the Forest of Dean, the death of a barge-master from cholera during the epidemic of 1831-2. Once, describing an atrocious burglary which culminated in attempted murder and the blinding of one of the victims, he told Salt he had subscribed a sovereign of the company's money to the relief fund. Another time, he wrote thanking Salt for a Christmas barrel of oysters. The series is not now complete, but Denyer's surviving letters cover eighteen of the thirty-two years 1811-1842, with the result that the records of the Thames & Severn Canal are singularly rich in human and technical interest.

Quite early in this period, in the summer of 1815, Shelley, Mary Godwin and two friends arrived at Inglesham with the idea of crossing to the Severn and rowing their wherry as far north as they could get, but their airy dreams were quickly dispersed when they found that use of the canal would cost them £20,[4] just as much as if they were the crew of a fully laden 60-ton barge. The incident was typical of the management's attitude to tourists for more than a century, ever since 1794 indeed when the committee decreed that no pleasure boat should be allowed to use the canal unless one of their members was aboard or had given written permission. Even when the Thames Navigation Committee of the Corporation of London arrived at Brimscombe in a shallop in the summer of 1816, they only escaped payment by telling Denyer that they had "passed toll free upon every canal" they had so far visited.

The City Committee, who had recently spent large sums improving the river below Staines, were trying to find out how far trade passing from the western canals to the Thames was hindered by the condition of the river under the commissioners' jurisdiction. Their subsequent report showed of course that the upper Thames was very imperfect, but it showed also that the Thames & Severn Canal was then "defective on account of the weeds" and "the decayed state of some of the lock gates".[5] Even with their finances as straitened as they were then, it was inexcusable of the company to add leaking lock gates to the many causes of loss which plagued them: such as evaporation, particularly severe in open country like the Cotswolds; the depredations of vermin, which, in spite of employing rat and mole catchers, perforated the banks; damage by cattle around the statutory drinking places; the effect of the unequal lock falls; and, most serious of all, the effect of the geological formation of the Cotswolds.

East of the tunnel, where the summit was cut through the great oolite, no puddle was proof against the action of the springs which in wet weather welled up through the fissures in the rock, perforated the lining, and left holes through which water escaped in increasing quantity as the level of the water table fell in the surrounding rock. Time and again leakage broke out in the neighbourhood of Coates, Trewsbury, Thames Head, Smerrill and Blue House, over lengths extending from a hundred yards to half-a-mile. Eight of the weekly returns from Siddington in 1809 referred to separate leaks which Proctor had done his best to staunch. In January 1817, Charles Jones discovered forty or fifty holes near Coates, and when one of these, described as "about the size of a common hat", was under repair, the men traced the fissure three feet underground to a space large enough "to hold a man". Again, in January 1831 "a tremendous leakage broke out" at Blue House and "lowered the Summit 5 inches" before it was temporarily plugged. Under such conditions, the summit more nearly resembled a sieve than the "Capital Reservoir" Robert Whitworth had intended it to be. Maintenance of the designed depth of six feet so soon proved impossible that by 1797 the gauge weirs had been lowered to 5 feet 4 inches. They were lowered still further subsequently: by 1830 the maximum depth had long been 5 feet; from 1879, if not some years before, it was 4 feet 3 inches or 4 feet 6 inches. Only meticulous examination and repair could maintain even the reduced depth, for the effect of leakage was cumulative, the escaping water exposing the clay lining to drying winds and hot sun which cracked it so that more leaks formed. Deterioration was therefore very rapid when times were hard and maintenance was neglected.

Where locks were many, the effects of a leak were contained within reasonable bounds, but long level pounds had to be subdivided, and any specially vulnerable spot such as an aqueduct or embankment protected, if possible, by devices closing automatically. Brindley tried doors hinged in the

Circular gauge weir at Bowbridge, near Stroud, in 1950.

bed of the canal and raised from horizontal to vertical by any unwonted flow of water, but it was soon found that damage by passing boats and deposit of mud made these unreliable.[6] A single gate or a pair of gates, as hung in a lock, was supposed to be similarly self-acting. Five of this type were set up in the summit level of the Thames & Severn from the start, notably east and west of the tunnel and at each end of Smerrill Aqueduct, but although they continued to be fitted much later elsewhere (as on the Birmingham & Liverpool Junction Canal and at Saul), both Clowes and Denyer considered them perfectly useless, as they certainly were against a leak developing on their blind side. Denyer preferred the cheapest and simplest form of all, which, though neither self-acting nor even quickly operated, was effective against a leak on either side. This was the plank stop, a row of thick planks inserted one above the other and secured by wedges in vertical grooves cut in masonry abutments, often of existing bridges. Grooves had been provided at two points in the tunnel from the beginning. Others were cut in various bridges soon after. Then in 1817 all the gates except those at Smerrill were replaced by fittings for plank stops, and later their number was multiplied until there were not less than eighteen along the summit and two in the Five Mile Pound.

In spite of the comments of the City Committee, little but routine maintenance was done for several years, and by 1820 the canal was so short of water that the company asked the advice of the well-known contractors,

Hugh and David McIntosh. David inspected the line from Stroud to Latton, but not beyond, and from his valuable report derived an extensive programme of overhaul and improvement which, under the wise direction of Richards and the skill and energy of Denyer and Jones, was carried out during the following quarter century, raising the canal to a high state of efficiency.

Giving his attention first to the loss of water, McIntosh remarked that "the nature of the ground is such in some parts . . . that in spite of whatever repairs may be attempted leaks will occasionally break through", and advised that those places should be examined and repaired once or twice every year. Much of the remainder of the summit should, he recommended, be stripped and relined with the best clay to a depth of 2½ feet on the bottom and 3 feet on the sides. This was done, part by McIntosh under contract in the autumn of 1821 at a cost of £1,169, and more by direct labour under Denyer and Jones in the two following seasons. Some of McIntosh's work was spoilt by false economy, old side puddling being left in position where it was 'tolerably good' and subsequently separating from the new bottom which, according to Denyer, it was very prone to do; and some was ruined by carelessness, as when one of McIntosh's men left a wooden scoop buried in the puddling. Denyer was therefore probably justified in claiming that he and Jones did the work better and cheaper.

At the same time the locks were thoroughly overhauled, one being entirely rebuilt. More important still, in order to economise water at those wasteful locks whose height was greater than their neighbours', McIntosh recommended the provision of side-ponds, as extensively used on the Kennet & Avon and Grand Junction canals. These were wide shallow basins which took from a third to half the water when the lock was emptied and stored it until the chamber had to be refilled.[7] Side-ponds three feet deep were installed alongside the five upper locks at Daneway, probably in 1823, but the lock most in need of one, the lower of the two at Wilmoreway, was not dealt with until 1831. Built in expectation of the supply from the Boxwell Springs, Wilmoreway Lower had a fall of 11 feet, and standing between neighbours 7½ and 6 feet high, exhausted the upper pound and overfilled the lower every time it was used. Here Jones built a side-pond 2 feet 2 inches deep and connected to the chamber by a cast-iron pipe closed by a cover operated from above. Taking about two-fifths of the contents, this reduced the consumption of water to that of a lock 6 feet 7 inches high, so bringing Wilmoreway Lower more nearly into line with its neighbours. But the working of the sluices each time the lock was filled or emptied, first in the pipe and then in the gates, increased the time taken to pass through the lock from about 5½ to 9 minutes, and was not a matter to be left to an impatient boatman. So a new cottage was built alongside for the watchman who had previously been stationed at Cerney Wick.

Yet another of McIntosh's suggestions was to widen each lock so that

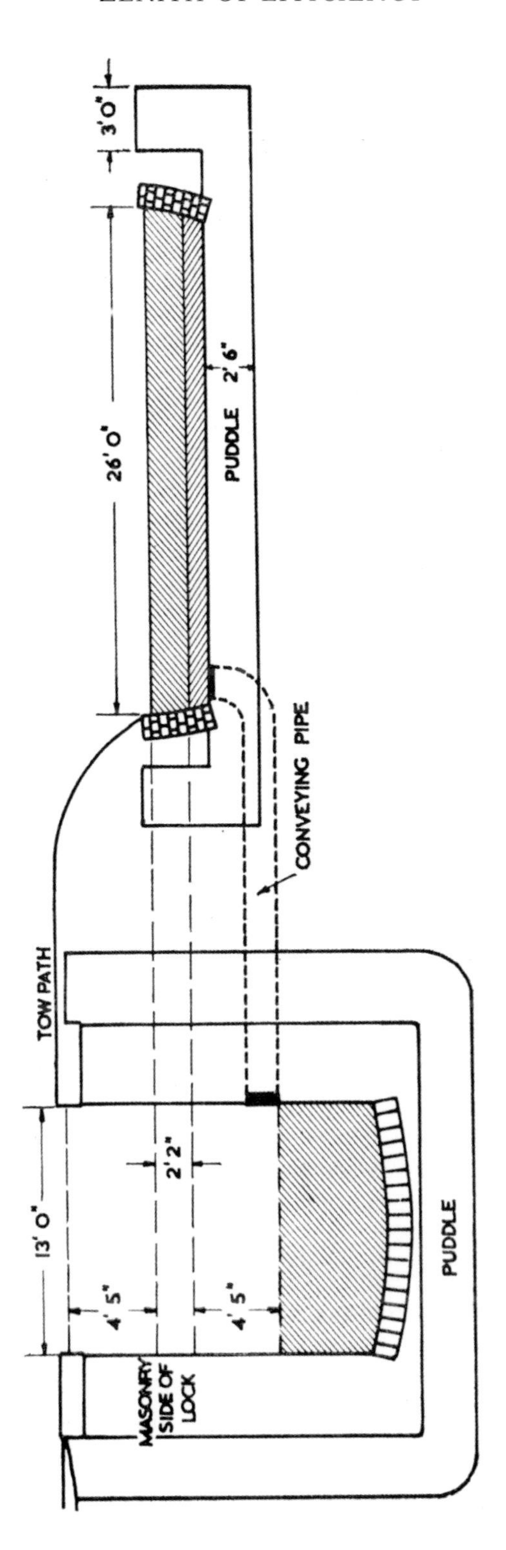

Charles Jone's design for a side-pond at Wilmoreway, 1830.

narrow boats could pass in pairs, and to hang a third pair of gates to reduce the effective length when they did so, but this entailed an expensive rebuilding of every one between Stroud and Latton and was never adopted. Instead, the committee developed a far simpler and cheaper method, but as it involved permanently shortening each lock and excluding Thames barges, which McIntosh's plan did not, they proceeded very cautiously. Doubting whether they had the right to take such a step, they first obtained authority by inserting in a Bill promoted by the Wilts & Berks in 1835 a clause enabling them to do so on the ground that the junction of the two canals was "much impeded" by the excessive consumption of water in the Thames & Severn locks; but even so they hesitated to use the facility for some time lest they should find themselves obliged to buy up "every vessel in the habit of navigating the Canal that would be prevented so doing by the alteration", and then shortened just one lock, at Latton at Easter 1837, in order to test the demand for compensation before being irretrievably committed. Another four years passed before the committee decided that as "locks . . . arranged for the accommodation of the largest sized barges navigating the Thames were no longer required owing to the discontinuence of those boats", it was safe to deal with the remainder. In 1841, seventeen locks in the Golden Valley, and in the following year seven at Siddington and South Cerney, were shortened about twenty feet in the same way as Latton, by throwing a masonry arch across the upper end of the chamber and re-hanging the gates, reducing the consumption of water by about 20 per cent. The cost was £1,565 16s. 6d., an average of £65 apiece. It seems probable that Denyer and Jones used the opportunity to equalise the height of several of these locks, for the seven leading to the summit at Daneway, four of which previously rose 8 feet and three 9 feet, were subsequently 8 feet 5 inches each, and the two at Wilmoreway were 9 feet 3 inches each instead of 7 feet 6 inches and 11 feet. The shortening proved entirely satisfactory until the final decadent years, when the water level often fell so low that, as the lock was filled, air was trapped and compressed below the crown of the arch, blowing it up.

Anxious to improve the working of the locks, Jones wanted to fit new quick-acting paddles, so a pair of fan-shaped discharging paddles of a design tried successfully on the Sankey Navigation was cast by a local smith for £14 5s. and fitted in new lower gates at Cerney Upper Lock. These emptied it in little more than two minutes, a great improvement on the five or more minutes of the original type, which moreover worked ever more slowly against masonry encrusted with age. A second type was a cast-iron, horizontal slide paddle working in an iron frame and tried in the upper gates of one of the Whitehall locks, which it filled in 2¼ minutes, and furthermore discharged the water vertically so that it did not rock the waiting boat. Jones wanted to fit both types throughout, but unfortunately Richards and Denyer, by then

Humpback Lock, ratchet and pawl for raising the paddle, in 1949.

apprehensive of railway competition, felt the cost was not justified.

The extensive works of the early twenties had so successfully conserved water that for a time the committee, over-optimistic as usual, hoped to cut out either the Churn Feeder or the Thames Head engine. The more practical Denyer was unconvinced, and sure enough by 1830 the position, aggravated by the increasing volume of trade, was again desperately acute. All sorts of expensive plans were debated by the committee: another engine to pump from the Thames? or the Churn? or from wells at Siddington? or at Blue House? a feeder from the Coln? a new line at a lower level to replace part of the summit? It is possible that they even discussed calling in a water-diviner, as in one of his letters, the chairman remarked that he liked "a recommendation above ground better than the old one of sending a Staffordshire conjuror to tell where water is to be found below the surface". Denyer and Jones, however, knowing that the Thames Head engine was capable of pumping double the quantity of water available in summer, saw that the solution lay in increasing the supply to the well. This they believed could be done by tapping nearby springs which were known to be unaffected by the working of the engine. In 1820, an unusually dry summer gave them the first opportunity for twenty-two years to examine the underground galleries, and they were so impressed by what they saw that Denyer decided to make the maximum use of so fleeting a chance, and on his own responsibility ordered a resumption of work on the 'Little Tunnel' driven by Philps at the close of the eighteenth century. The committee condoned his action even if they did not approve it, and when the next opportunity arose, in August 1827, Denyer was a great deal more sure of his ground., He ordered miners to extend the tunnel, set about deepening the pit sunk experimentally at The Firs, 500 yards away, in 1790, and sank another between there and the pumping station. Finding that the two pits were related and promised "*a very great increase*" of the supply, he decided to speed up the work by excavating an adit instead of driving a tunnel, afterwards roofing it with stone and restoring the top-soil.

He had not gone far, however, before Robert Gordon, the arrogant squire of Kemble, ordered the work to stop and appealed to the commissioners established under the Act of 1783, claiming compensation not only for his tenant (which the company conceded) but also for the villagers of Kemble, Coates and Tarlton, whose wells, he alleged, had been impaired by the Thames Head pump and ought therefore to be deepened at the company's expense. Faced by this threat, the proprietors consulted H. R. Palmer, Vice-President of the Institution of Civil Engineers, who, after visiting Thames Head, advised them that their action was justified, that any effect upon wells would be very difficult to prove, and that a claim for damages arising from thirty years' usage was unlikely to be sustained. Nevertheless, the commissioners did in fact find for Gordon, but as they made no award, the company were faced with the alternatives of negotiating with the squire or continuing the excavation and leaving him to seek his remedy at law.

Perhaps John Disney had lost interest, certainly he had lost his earlier drive, for although the need of water grew even greater, nothing was done until he was succeeded by W. P. Richards. Deciding to try personal negotiation, Richards found he was courteously received by Gordon and came away "much gratified" by their discussion it was not the only time his judgment was affected by plausibility and charm. Not until nine months later did the squire consent to withdraw his opposition, and then only on condition the company would pay him £1,500 and give up their right to search for water on his lands. Richards recommended acceptance of these terms, but the proprietors considered the sum "much beyond any injury" the Kemble estate was likely to suffer, and resolved to appeal to arbitration. Richards was piqued, but Gordon, probably concluding that this course would be less profitable, lowered his figure to £1,000 and withdrew his condition, and on these terms an agreement was signed in July 1832. As during the development of the dispute both parties lost sight of the unfortunate tenant's rights, it may be doubted how much, if any, of the compensation ever found its way into his pockets.[8] Certainly, a month later, Gordon, then in residence at Leweston, Dorset, had not bothered to notify his Kemble steward that an agreement had been reached. Certainly he was a rapacious man, for when the Cheltenham & Great Western Union Railway was promoted four years later, he rooked them of £7,500 and obliged them to build a covered way through the grounds of Kemble House, to rebuild a bridge across the Thames, and to accept a prohibition against building any station on his estate.[9]

And so after nearly five years, Denyer was able to resume work on the adit. Unlike driving the 'Little Tunnel', excavation of the adit was not confined to seasons of exceptional drought, and though liable to interruption in very wet weather or delay when local labourers were harvesting, it went on year after year from July or August to November while the springs were at their lowest. Cutting through 'stiff clay blasting rock, quarrying blocks of stone up to half a ton in weight, labourers and Forest miners carried it 556 yards from the junction with the tunnel (at a depth of 23 feet) to The Firs, tapping fresh springs by drilling the rock with 3½ and 6 inch bits as they went. Even at the end of the second season, Jones reported that at least two-thirds of the water pumped by the engine was being drawn from the new works, and by August 1838 the yield so greatly exceeded expectations that the average depth in the well was 20 feet (rising to 33 feet or more in winter, falling to 10 feet or less in the dry season) and it was unnecessary to do more, The excavation cost less than £600, and so, even including Gordon's exaction, proved far less expensive than any of the alternatives examined by the committee.

But where is the place they called The Firs? That it is near Lyd Well, "held by some to be a Roman well and mentioned in Doomsday Book . . ., the last [site] at which water would fail in the valley", cannot be doubted, but although "there are indications that a culvert ran from this well for some 500 yards up the valley" to one of the secondary wells of the pumping station, no one — not even John Taunton in 1882 — has been able to do more than suggest that Lyd Well and The Firs are really one.[10]

While the adit was being excavated, the Boulton & Watt engine was thoroughly overhauled by a local engineer, John Ferrabee of Thrupp Mill, Stroud. It was none too soon, for many of the parts, then forty years old, were so badly worn that a breakdown was imminent; furthermore, though it was known that one boiler was worn out, examination revealed that the other also was on the point of failure. Ferrabee installed two new wagon-shaped boilers, supplied by the Cinderford Iron Company, improved the grates and flues, raised the stack, modified the cylinder to reduce steam consumption, added grease cocks and a 'Clock' to record the number of strokes. All this, completed in the summer of 1833, occupied some eight months and cost about £1,200, but it improved the output by 71 per cent and even showed a slight increase on the work done when the engine was new. It is possible that the ageing Thomas Toward felt the condition of the engine to be a reflection on himself (although Ferrabee was careful to refute this) and probable that he disliked the innovations, but he chose this time to give notice "after fulfillin a Sarvitude of 42 years as the Local Engineer". Denyer replaced him, without difficulty and far more cheaply, by Chivers, an engineman from the Forest of Dean, who for 26s. a week and with the aid of his son, took on the whole of the work previously done by Toward with the help of a firemen and a third man engaged when the engine was working day and night.

Chivers introduced another improvement, the cataract valve which had been fitted to the mine pumping engines he had known in the Forest. This was an ingenious, "but simple, device to control the engine when shortage of water in the well made it unnecessary indeed positively dangerous to work at full power. It involved fitting a special pivoted lever, which could be connected or disconnected at will, to operate the cold water injection cock of the condenser. The lever was weighted at one end in order to keep the cock closed unless a cup at the opposite end contained enough water to overbalance the weight. By regulating the rate at which water flowed into the cup, the engineman was able to reduce the speed of the engine as much below the normal rate of ten strokes a minute as he thought fit, instead of having to control the engine by hand and wastefully limit its stroke to as little as three feet, or perhaps even return to the well some of the water already lifted. This idea and his excellent maintenance of the engine brought him a rise of 9s. a week, but nevertheless he stayed little more than four years. Two further improvements followed: insulation of the cylinder and steam pipes by wooden casing filled with sawdust; and lagging the boiler, first with "Borradaile's Patent Felt" which proved to have little effect, and later and more successfully with ashes.

The depth of the navigation was improved not only by increased pumping from the well, but also by a systematic cleaning of the bottom, silted up over the years with the deposit from the direct flowing feeding streams.[11] Although some silt had been removed from time to time, by 1830 there was so much mud and so little water that narrow boats cut a furrow along the bottom and left a ridge on either side, making it almost impossible for the wider trows to pass. His first remonstrance being ignored, Jones wrote to the committee, describing dramatically the common occurrence of "horses straining their utmost with block tackle and ropes breaking to get the boats through the mud; at the same time Bargemen cursing and abusing the canal in every form". This took effect, and work was begun in 1832. Although a 'spoon' or scoop dredge, formed of a leather bag on an iron ring at the end of a long pole, lowered over the side of a boat, dragged along the bottom and raised by a windlass,[12] was used on the Thames & Severn in later years, the company had no such apparatus then, nor were they inclined to risk using any kind of dredger on those parts with the delicate clay lining. Jones was therefore obliged to do what he expressively called 'mudding' by hand, but the result was so effective that he was authorised to continue during normal stoppages until the whole canal should be clear. 'Mudding' therefore went on for many years, the climax coming in 1844 when Brimscombe Basin and Ham Mill pound were cleared under contract and "the quantity of mud taken out, computed at about 45 Thousand Tons", so far surpassed expectation that the contractor's estimate of £203 was exceeded by £90 and the committee felt constrained to allow him £60 extra.

Yet another long term improvement was begun by Jones in 1840. To protect the summit from the drying action of sun and wind in the season of short water, he drove a slanting line of elm planks, grooved and interlocked with a slip of wood, and filled in behind with "a loamy gravel puddle" which ensured that the underlying clay retained its moisture. The first length of 110 yards at Blue House cost nearly 10s. a yard, almost double his estimate, but was so successful that the committee authorised him to extend the 'plank piling' up to a limit of £150 a year, which he continued to do for some five years, gradually reducing the cost to 5s. 2¼d. a yard.[13]

In addition to these major works of improvement, there was a great deal of routine maintenance. In the tunnel, leaks were frequently stopped, distorted side-walling and invert intermittently rebuilt. Old lock gates had to be repaired, new ones made and hung, eight pairs in the one season of 1822, one pair made in haste in a fortnight when a boat breached Bowbridge Lock and the rush of water totally destroyed the gates. Decaying or damaged bridges of timber, stone or brick needed periodic attention; but a farcical situation arose at one, the accommodation bridge at Griffin's Mill, a mile below Brimscombe, which had been misused and damaged by a tenant who carted heavy loads of timber. After being neglected for nearly three years, while company and landowner wrangled interminably over responsibility for restoration, it fell into the canal, and only a complaint of loss of right of way eventually induced the company to put up in 1842 a 'temporary' footbridge — which appears to have lasted until rebuilt with lattice iron girders in 1903. Punts for carpenters and other maintenance men were built from time to time, for example one 50 feet long and 10 feet wide, with a cabin, made at Brimscombe in 1845 at a cost of £94 9s. 10d.

As far as possible, ample notice was given of impending stoppage for repair; and then in the last preceding days, the canal appeared unwontedly busy as boats with urgent cargoes crowded along it. Aware of the serious dislocation caused by repeated stoppages on contiguous canals, Denyer proposed in 1824 that the Thames & Severn, Stroudwater, and Wilts & Berks should all three do their annual repairs at Whitsun, a period favoured also by the Kennet & Avon and others. Synchronisation was not easy to achieve, however; some canals needed longer than others; the sixty or so labourers were not always obtainable ("scorching days" in May 1833 led to the early gathering of the hay crop); nor were they always, as Denyer put it, "in a proper state of sobriety" for work during the holiday period. He succeeded completely in 1827, partially in 1834; but in 1831 the Wilts & Berks chose Easter; and in 1832, although both the Wilts & Berks and Stroudwater needed three weeks for repairs, they managed to get out of step with each other as well as with the Thames & Severn.

Further dislocation was caused by fog, flood or drought on the rivers, and frost, to which all canals were very susceptible and the Thames & Severn especially so above Brimscombe and on the uplands. Ice-boats, 24 feet long, 8 feet wide and drawing 2 feet 9 inches, were built at The Bourn Yard before the close of the eighteenth century and probably conformed to the common type used on most canals and certainly on the Thames & Severn in its last years of operation. They were sturdy boats drawn by a team of horses and manned by a crew who clung to a handrail and rocked the vessel to break up the ice abeam and widen the passage already cut, but in severe winters their work was repeatedly undone. In 1830, for example, the canal was frozen over by 2 January, and though freed in stages from Brimscombe to Inglesham, soon froze once more. On the 22nd, under Denyer's supervision, a passage through ice nine inches thick was cleared as far as the tunnel, but it was blocked again within three days. The following week Jones set off eastward from Siddington, and increasing the number of horses to four, five and seven on successive days, he managed to reach Kempsford before sticking fast until the addition of another two horses freed the boat "with the greatest difficulty" and drew it on to Inglesham. Then frost descended again, binding the canal below Brimscombe as well as above. When the final thaw set in on 7 February, the whole line was cleared, so Denyer claimed, "many days before most other canals", but 'vast masses' of broken ice still remained.

The progress of this battle with the ice was characteristic of the assiduous devotion shown by both men, who never spared themselves in the performance of their duties, riding hither and thither on horseback "devising and Superintending the various works and Improvements". Denyer had no engineering training — the new devices were all produced by Jones — but he had learnt a lot from Jones: furthermore, the offspring of Father Thames and Madam Sabrina was herself an exacting teacher, ruthless in exposing ignorance, carelessness, or over-confidence. Denyer's acquired knowledge and the obviously confident relationship established between himself and Jones led to a satisfactory arrangement conserving the time and energy of both, Denyer supervising almost all works in progress west of the tunnel, Jones those of the summit level and the east. How effective their improvements were is proved by the average annual payments for water taken from the

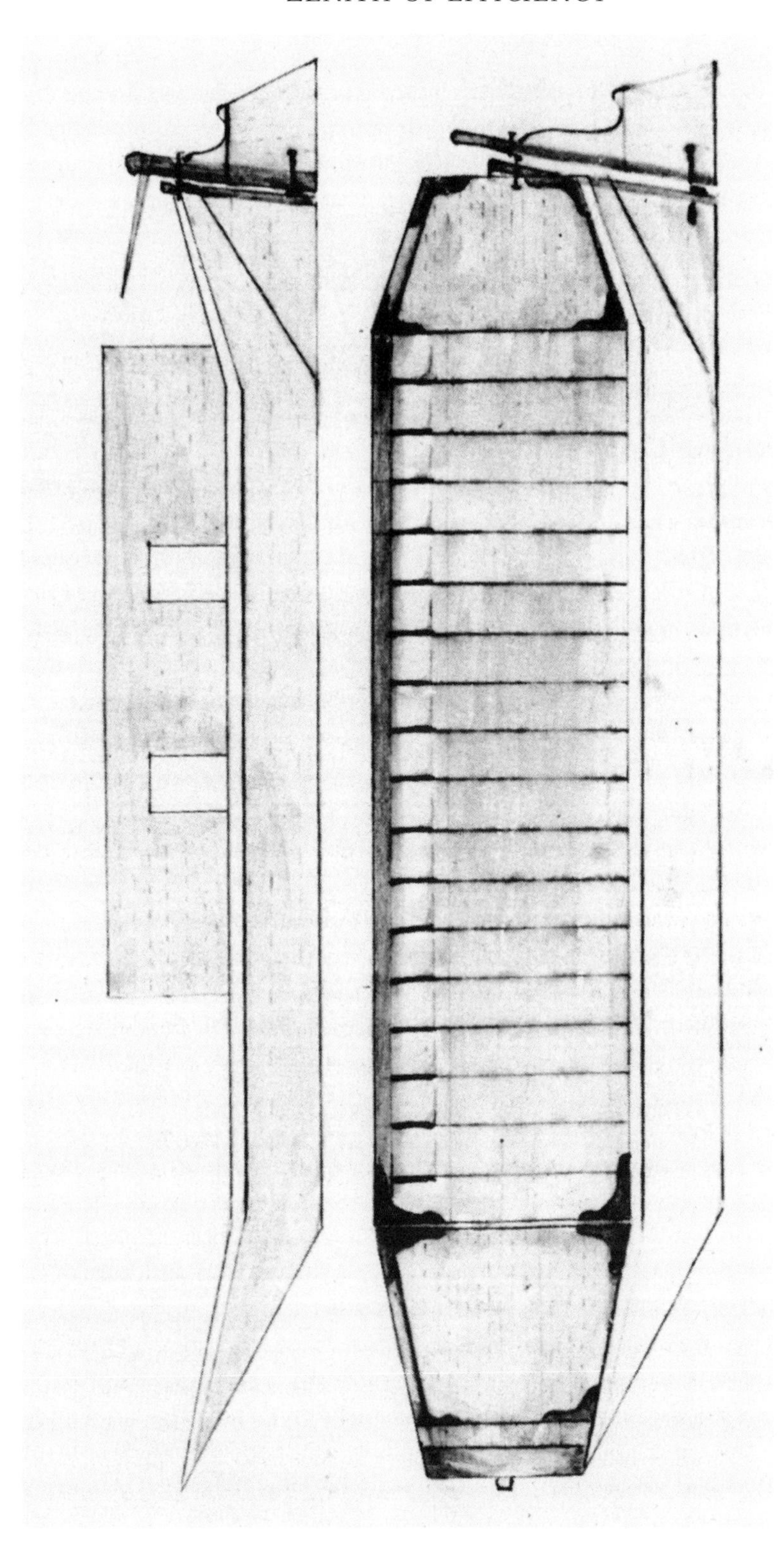

Drawings of the carpenter's punt at Brimscombe in 1845.

Churn, which fell from £355 in the twenty years preceding 1836, to £293 in the following decade, despite the great increase of traffic, which in that period reached its peak. Moreover, as the proprietors gratefully acknowledged, they had done all this "in addition to the Ordinary Duties devolving upon them" as agents respectively at Brimscombe and Siddington; Jones indeed, frequently away from home for days on end, often found it impossible to fulfil both rôles efficiently. In 1842, on the committee's recommendation, the proprietors decided to divide between the two men 7½ per cent of the profits remaining after distribution of 30s. on the Red and Black shares, which, when the year's accounts had been closed, amounted to £98 0s. 3d.

His own and other letters in the files show that as Denyer grew older he became masterful, pompous and overbearing, irascible and quarrelsome, but that to the end he served the company with shining devotion and integrity. On 27 December 1842, after only a few days' illness, he died from an attack of erysipelas. What did he look like? He was 62 years old, and his request made, but not granted, eight years previously "for a Gig in consideration of the difficulty he finds in riding" suggests that he had grown portly, but at the date of his death photography was not yet in general use, and though he certainly did not lack the self-esteem to have his portrait painted, none can be traced.

Within a month his successor, William Willson, was appointed at a salary of £200 a year with house, garden and lands. Charles Jones, who had applied, was asked to remain as clerk-of-the-works, assuming the responsibility for maintenance throughout the line, and offered the same salary, £200 a year, with £35 for extra expenses. The arrangement did not work out as intended, for Willson had his own ideas. He put a stop to Jones's extension of the plank piling, preferring, probably unwisely, a new method of puddling. Within a short time it was reported that Jones was "continually opposed to the directions given him by Mr Willson" and that a "want of cordiallity" had developed. Enthusiastic and enterprising, but plausible and lacking integrity, Willson had the complete confidence of the committee, who during this crucial period were deprived of the wisdom and experience of John Stevenson Salt whose judgement of character would never have been so faulty. Jones found the committee ranged against him and his salary reduced. The long intimate partnership of the leading figures had indeed been broken. The scanty correspondence of the succeeding years is dull, correct and formal.

Nevertheless, Willson was progressive, and he was soon off to Oxford, Staffordshire, Berkshire and Somerset in search of new ideas, returning with several. One was for the introduction of steam tugs similar to some he had seen on the Birmingham & Liverpool Junction Canal. The committee approved this idea, but doubting whether the natural ventilation of the tunnel would dispel the smoke, consulted William Cubitt, who was apparently unable to reassure them. Another, however, resulted in the canal acquiring a rare and distinctive feature, the "Machine for weighing loaded Boats" at Brimscombe.

Hitherto it had been the practice to check false declarations of cargo by 'gauging boats': fixing metal plates on either side of bow and stern of an empty vessel and recording in a book the height of the plates above the water line as she was loaded with known weights.[14] Obviously each boat using the line had to be so gauged and recorded, copies of the record distributed to each toll house and kept up-to-date, and the tedious process of measuring the depth, consulting the book, and calculating the weight of cargo performed on occasions of doubt. Equally obviously it was far easier to accept the master's word. In any case, with cargoes of coal it was customary to charge toll on tickets of weight issued by the collieries, and it was well-known that through the connivance of those concerned masters frequently carried several tons more than they paid toll on.[15] One way and another, abuse could not be prevented, and as an "infinite variety of different shaped & sized vessels" used the Thames & Severn Canal, prevention of fraud was more than ordinarily difficult and there was a probability that an elaborate boat weighing machine would justify its cost. There is record of very few similar machines in the country: a weighing house on the Grand Junction was probably no more than a gauging dock; but at Cardiff, the Glamorganshire Canal had a boat weighing machine made in 1836 and now preserved alongside the Waterways Museum at Stoke Bruerne in Northamptonshire;[16] there was possibly one on the Monmouthshire Canal;[17] and there was one on the Somersetshire Coal Canal at Midford, which was that seen by Willson. Having submitted a drawing and later a model, he persuaded the committee to adopt his plan "with all proper expedition" and so "check the frauds known to be practised by the Boatowners".

The machine was an adaptation of the compound lever weighbridge invented a century earlier by John Wyatt of Birmingham. To house it, a small dock was formed alongside the wharf in front of the transit shed, with gates at each end, and three cast-iron columns on either side supporting roof and superstructure. When the gates had been closed and the water run off by gravity, the vessel lay on a cast-iron cradle slung by five radial rods from each corner of a massive frame resting on the superstructure; and the weight was then taken on the weighbeam so poised on a fulcrum that the leverage exerted was 93 1/3 to 1, and the weight proportional to one ton was no more than 24 lb.[18] The estimated cost was £1,062 5s. 11d. Results fully justified the expense: in twelve months from February 1845, 1,267 tons of overweight were detected on 688 of the vessels weighed, and the yield of £153 18s. 4d. returned more than 14 per cent of the cost. Nor did it then prove merely a deterrent, as in the next four months extra toll was collected at a slightly increased rate. The boat weighing machine was dismantled in June 1937,[19] and the dock converted into a swimming-pool quickly condemned by the health authorities, but Willson's model has survived

Model of the 'Machine for weighing loaded Boats' at Brimscombe Port, made in 1843 for the committee's approval.

Dismantling the boat weighing machine in Brimscombe Port in June 1937. This had been erected in 1845 to check frauds known to be practised by boatmen avoiding the payment of tolls.

(*Photo courtesy Stroud News & Journal*)

and is now preserved in the Gloucester Folk Museum.

It was several years before the indulgent committee realised — or would admit — just how dishonest Willson was. Nearly three years after his appointment, they learnt that he had been insolvent before joining them, but far from letting that affect their confidence, they acknowledged their appreciation of his "valuable services . . . and fidelity" by giving him a rise of £50. In the summer of 1847, however, they found his accounts deficient by more than £1,000. Nevertheless they treated him with extraordinary lenience, deciding that it was "to the interest of the Company" to retain his services. Even when he later confessed that his liability was double that amount, they refused to accept his resignation, though they did instruct him to sell his effects and repay them in instalments of £50. Not until August 1852, when examination of his accounts showed that he had continued to pocket rents and cash payments and that his total liability was £3,090, did they admit 'the entire destruction of their confidence' and the necessity of dismissing him.

His successor, appointed the same month at a salary of £250 a year, was John Hooke Taunton, without doubt the most distinguished and highly qualified officer directly employed by the company. Born in Devon, educated at a private school, apprenticed to a firm of civil engineers, by the age of thirty-one he was already an Associate of the Institution of Civil Engineers and less than six years later became a full member.[20] In appointing him to a position whose responsibilities hitherto had been commercial rather than engineering, the committee probably bore in mind that Charles Jones's retirement could not be long delayed, and as for the commercial side, the "very strong testimonials and recommendations" they had received would certainly have revealed that Taunton had shrewd business acumen.

As a pupil, he had shown an exceptional talent for engineering drawing; as a surveyor, he had won the high opinion of two distinguished engineers, J. M. Rendel and I. K. Brunel; and to his skill in these two directions, we owe the exquisite series of plans and sections he afterwards produced of Sapperton Tunnel and the whole line of the Thames & Severn. Two years before going to Brimscombe, he had been extending the Reading waterworks and had successfully forestalled the challenge of a rival company by buying up "all the bricks available in the neighbourhood" and pushing ahead with the building of a reservoir although parliamentary authority had not then been obtained.[21] No doubt the canal proprietors were delighted to obtain the services of so able a man at so meagre a salary — less than had been paid to Clowes seventy years before — and so agreed the more readily to his practising privately as a consulting engineer, which he did thenceforward, generally in connection with waterworks.

The committee appointed him, as they had his predecessors, to the position of 'Agent at Brimscombe Port', but the bond of indemnity each had endorsed used the form 'Clerk or Manager' or, in Taunton's case, 'Agent or Manager', which in fact described their function better. Subsequently, however, Taunton used a style of greater dignity to which he alone of them was entitled, that of 'Engineer and Manager'.[22]

At Brimscombe, the matter most urgently needing Taunton's attention was how to keep the Thames Head pump going? Twenty years had passed since Ferrabee's overhaul of the Boulton & Watt engine, and more than three years earlier Willson had warned the committee of its "decayed and worn out state". When Taunton examined it, he reckoned that £800 to £1,000 would have to be spent on reconstruction. In his opinion, this would not be justified, as although replacement would be more expensive, a Cornish engine using steam at higher pressure and working expansively would be so much more economical that there would be a substantial saving over and above the interest on the capital expended.

In October 1853, the committee authorised him to go ahead, and in little over six months, Taunton had found, bought, and installed a second-hand engine built in 1852 by Thomas & Co. of the Charlestown Foundry for Wheal Tremar, near Liskeard, and scarcely used before the mine closed down. Dismantled by Thomas James, a well-known Cornish engine builder who subsequently re-erected them, the parts were shipped from Looe, followed by a new boiler and pumpwork from Sandys, Vivian & Co. of Hayle. On 8 February 1854, the day after the vessel had arrived in the Severn, dismantling of the Boulton & Watt engine began, and as the new engine was set to work on 28 April, the pump was out of action for little more than eleven weeks, and that in the season when it was not normally required. The total cost, including enlargement of the boiler house, was £2,497 15s. 10d., less £135 for the sale of 'old iron' the ungracious epitaph of the Boulton & Watt engine.

Although the new boilers had a working pressure of 27 lb to the square inch, in practice about 20 lb was maintained, feeding a cylinder 40 inches in diameter and driving a piston through a stroke of 9 feet eight times a minute. Like the Boulton & Watt, it was a beam engine working a bucket pump, but in a larger barrel, 30 inches in diameter. With grates burning coal of a cheaper quality, a cylinder in which one-third of the stroke was performed by expansion of the steam, and a pump which raised nearly 50 per cent more water at each stroke, the new engine did more than three times the work of the old at little more than a quarter of the cost. Working day and night, it delivered about three million gallons in 24 hours. It remained at Thames Head long after it had ceased to be used, indeed until the drive for scrap metal of all kinds early in the Second World War.

This was the last of the great works of improvement undertaken to make the canal, as Richards had written to Denyer twenty years earlier, "as perfect as possible" to meet railway competition. Many were financed out of revenue, but five or six costing £1,000 or more were covered by withdrawals from the contingency fund. This fund had been formed after the financial reconstruction of 1809 to receive the dividends withheld from those holders of Black shares who had not subscribed to the half-share issue in 1796. The liability had been extinguished in 1829, but the fund had been increased by withholding all dividends for the three years 1821-1823 at the start of the programme of improvements, and later by the balance of profits after paying 30s. on all shares in 1831-1836, the committee having decided that no distribution should exceed that amount until the contingency fund reached £5,000. There had also been another substantial contribution, mentioned in the next chapter, so the fund was by no means exhausted after buying the Cornish engine and writing off Willson's debt.

REFERENCES

1. DNB John Disney 1779-1857:
2. Information kindly supplied by the Clerk of the Goldsmiths' Company.
3. Sutcliffe, op cit, 1816, p. 128.
4. Dowden, E., *The Life of Percy Bysshe Shelley*, 1886, pp. 526-9.
5. The Librarian of the Port of London Authority kindly supplied a transcript of the relevant part of the Committee's report.
6. Details of the various forms of stop and their operation are given in: *The History of Inland Navigations* 1766, pp. 47, 53; Young, A., *A Six Months Tour through the North of England*, 1771, vol. 3, pp. 208, 216-8; Phillips op cit, p. 98; *Edinburgh Encyclopaedia* vol. 15 part 1, 1830, p. 315; Cresy, E., *Encyclopaedia of Civil Engineering* 1856 ed., vol. 2, p. 1557; Stevenson op cit 1886, p. 20.
7. For side-ponds see: Sutcliffe op cit, pp. 106-7; Rees, *Cyclopaedia*, Canal; *Encyclopaedia Britannica* fourth ed., 1810, vol. 5, p. 111; Dupin op cit, vol. 1, p. 295; Priestley op cit, pp. 334, 388.
8. Cf Rudge, T., *General View of the Agriculture of the County of Gloucester*, 1807, p. 337. Rudge remarked that when canals were cut, the tenant's loss of land "is seldom estimated at its real value: a high price is given to satisfy the owner, while the only satisfaction made to the late occupier, whose estate perhaps is separated inconveniently into two parts, is a proportionate reduction in rent".
9. MacDermot, E. T., op cit, vol. 1, pp. 163-5. The present Kemble station was built in 1882.
10. Richardson op cit, 1930, p. 106. The six-inch map of 1884 and later 25 inch maps in the British Museum show the three wells at the Thames Head pumping station and a spring and pump at what must be the site of

Lyd Well, but neither Lyd well nor The Firs is named.

11. The French had found out the unfortunate effect of direct feed from running streams soon after completion of the Languedoc Canal — see Frisi op cit, 1762, p. 157, and Vallancey op cit, 1763, p. 116.
12. For the spoon dredger, see Weale's *Quarterly Papers on Engineering* vol. 1, and Stevenson op cit, pp. 227-8. Descriptions of the spoon dredger used later on the Thames & Severn were given me by the late Billy Stadden, sometime boatman and canal employee, and by the late C. W. Hawkes, sometime Principal of Brimscombe Polytechnic.
13. 'Pile-planking' of the type driven by Jones is described in detail in Rees, *Cyclopaedia*, Canal.
14. The process of gauging is also fully described in Rees, *Cyclopaedia, Canal.*
15. The late P. G. Snape, Clerk to the Stroudwater Navigation, told me that the practice of loading boats at the collieries with several tons more than the weight shown on the ticket was common enough in his time; and see Randall, Broseley, 1879, p. 166.
16. Information kindly supplied by the Curator of the Waterways Museum, Stoke Bruerne, Northants.
17. Dupin op cit, vol. 2, p. 329.
18. The late C. W. Hawkes of Brimscombe gave me this and other information about the boat weighing machine.
19. *Stroud News*, 11 June 1937.
20. *Proceedings of the Institution of Civil Engineers* vol. 112, 1893, pp. 360-4, obituary of John Hooke Taunton.
21. Ibid.
22. I have been unable to trace any official confirmation of Taunton's change of style, but he used that of Engineer and Manager when giving evidence before the Select Committee of the House of Commons on the Thames & Severn Canal Navigation Railways Bill on 16 April 1866 — which was eight months after the retirement of Charles Jones.

CHAPTER TWELVE

"LAST OF THE RAILWAY PRINCES"

The threat of railway competition, apparent ever since issue of the Great Western's first prospectus in 1833, became more menacing with the promotion on 13 October 1835 of the Cheltenham & Great Western Union from Swindon through Stroud and Gloucester. Within a fortnight, one of the railway directors had dropped a hint that his company might buy the canal, and the canal proprietors had readily decided to negotiate. But soon finding that the suggestion lacked substance, and probably hoping to force the pace, they decided to promote a railway of their own between Stroud and Cirencester.[1]

Their anxiety to come to terms before any railway was even built was shared by many other canals. In the years 1845-1847, the years of the Railway Mania, so many surrendered on the best terms they could get that about a quarter of the total canal mileage passed under railway control in one way or another,[2] including the prosperous Birmingham and even the wealthy Trent & Mersey. They had good reason to do so, for there were many defects which the new form of transport did not share.[3] As the companies were not carriers, they were out-of-touch with public needs, reluctant to reduce tolls, unwilling to co-operate in the quotation of through tolls, uninterested in developing steam power; while the carriers were mostly men of little substance, unable to provide more than a minimum of vessels or make use of steam, reluctant to pay compensation for damage or pilfering of goods in transit.[4] Bad enough, but what made proprietors quail at the first hint of railway promotion was the knowledge of defects more obvious and far more deadly: the delays caused by long flights of locks, circuitous routes, tunnels without towing-paths; but, above all, the liability to total interruption for unpredictable periods by flooding of adjacent river navigations, by summer drought, winter frost, accident, and especially by the ordinary needs of maintenance, when renewal of puddle or the repair of even a single lock closed a canal to through traffic.[5] The records of the Thames & Severn Canal for eight years, five of them consecutive, reveal a staggering aggregate of days when navigation was wholly or partially stopped from these causes:[6]

Year	By frost	For maintenance	By other causes	Total
1809	21	16	14 Thames flood	51
1822	0	127	—	127
1826	20	26	70 Stroudwater-Berkeley Canal junction	116
1827	19	39 at least	—	58
1828	0	40	—	40
1829	4	63 Tunnel	—	67
1830	40	5	35 Stroudwater repairs	80
1839	0	28	—	28

It may be argued that the Thames & Severn was uncommonly susceptible to these influences, but whatever allowances are made for local conditions, there remains a clear implication that canal navigation in general was subject to interruption on an astounding scale. It is obvious that most merchants and traders would prefer to rely on a new system of transport once its regularity was proved rather than hold stocks to guard against emergencies.

When the Bill for the broad gauge Cheltenham & Great Western Union Railway was introduced in 1836, the Thames & Severn company were joined in opposition by a powerful ally, the London & Birmingham Railway, eager to promote a standard gauge line to Cheltenham by a route twenty miles shorter via Aylesbury and Oxford, which would not have competed with the canal. As this preliminary skirmish in the Battle of the Gauges was likely to be bitter and expensive, the Cheltenham & Great Western Union decided to buy off the opposition of the canal company with an offer of £7,500 "as a compensation for damage to be sustained". This was accepted, and the sum, paid in instalments over three years with interest at 5 per cent, was invested in the contingency fund. After a "long and arduous contest", the broad gauge railway obtained its Act, but, deserted by many of its Cheltenham promoters who yearned for the shorter route, it had great difficulty in raising money and had to abandon its initial plan for a gentle climb at 1 in 330 up the Golden Valley and a long deep-level tunnel at Sapperton, and adopt instead a ruling gradient of 1 in 60 cheaper to build but far more expensive to work. Even so its rails advanced but slowly. It was opened from Swindon to Cirencester on 31 May 1841, absorbed by the Great Western in 1843, and extended to Gloucester, whence the railway to Cheltenham was already open, on 12 May 1845.[7]

The railway was carried so high above the floor of the valley that it was ill-suited to serve the mills clustered beside the River Frome and the canal, and its opening therefore affected the long-distance traffic of the Thames & Severn, such as that carried by the Gloucester and London fly-boats, rather than the local trade. Carriage of railway materials had raised tonnage, receipts and profits to a peak in 1840-41, and the tonnage (though not the receipts, for the canal reduced many tolls in October 1844) to a second but lower peak in 1844-45. Disregarding the returns of these abnormally inflated years, and comparing the period 1846-54 with that of 1831-39 before the railway was built, the coal trade actually increased by 11 per cent, the total tonnage decreased by a fraction, but receipts fell by 22 per cent, reflecting not only the reduction of charges but also the loss of through trade. For sixteen years, dividends of 30s. or more, rising to a maximum of 62s. 6d., had been paid on all shares, but in 1847 holders of Black shares received only 21s. 6d.

Far more damaging was the effect of the South Wales Railway, opened from Gloucester to Chepstow in September 1851. Meeting on 8 January 1852, the proprietors of the Forest of Dean Railway approved the sale of their company to the South Wales Railway, and on 24 July 1854, the old tramway, re-aligned, was opened as a broad gauge railway.[8]

The effect was immediate: in the following decade the average annual import of coal to the canal was 19 per cent less than in the previous decade, and in 1865-1874 there was a further fall of 35 per cent. The last dividend on Black shares was 2s. in 1853. In 1854 profits were insufficient to pay the full 30s. on Red shares, and 1864 was the last in which any dividend was paid. No wonder the committee again entertained ideas of turning their canal into a railway.

Happily, few of those whose devoted service had raised the canal to its peak of efficiency were witnesses of its ruin. John Stevenson Salt died in London on 16 August 1845 at the age of seventy;[9] the proprietors recorded their "regret at the loss of their old and valued Friend". James Perry's eldest son, Thomas, ceased to be a member of the committee in 1847. John Lane last wrote the minutes in 1850. News of "the death of Mr Disney, an old and respected Member of this Committee. . . was received with much regret" in 1857: although as chairman of the Forest of Dean Railway, he had bowed to the changing times and signed the canal's death warrant, it was his earlier policy that had rescued the company from insolvency and laid the foundations for such prosperity as it achieved. William Parry Richards, whose failing health had prevented his regular attendance for several years, resigned on 21 April 1860 after occupying the chair for thirty-two years. In August 1865, Charles Jones, whose arduous duties had been beyond him for at least ten years, was retired on a pension of £90 a year, the largest the company ever granted, and lived between five and six years to enjoy it.

As the reins fell from the hands of the older generation, a younger Lane, a younger Disney, and William Perry Herrick (Thomas Perry's heir) continued the long association of the three families with the canal, but it was the three sons of John Stevenson Salt who increasingly dominated its affairs. John succeeded his father as treasurer, and both he and his elder brother Thomas each served a term as chairman, but it was the third, William, who played the most surprising part. In partnership with his brothers in the family

Silhouette of John Stevenson Salt shortly before his death in August 1845.

bank, William was best known as an antiquary specialising in the history of Staffordshire.[10] Joining the Thames & Severn committee two years after his father's death, William Salt then held 95 shares, some Black, some Red, and in spite of the falling receipts and unpromising future, he set out to increase his holding. Buying ever greater numbers as more and more became available after dividends had ceased, by 1862 he held 836. As proprietors attending meetings at this period represented, in person or by proxy, only about a thousand shares, William Salt was then in the position to ensure acceptance of whatever policy he had persuaded the committee to adopt, and he had achieved this ascendancy at modest expense, buying Blacks at £1 apiece and Reds at £10, so that the whole had been acquired for about £3,500 spread over fifteen years or so — a tithe of what he spent on his Staffordshire collection.

Now, as partner in Stevenson, Salt & Co., treasurers of the Trent & Mersey Canal, William Salt had seen, and very likely benefited from, remarkable dealings in that company's shares. The canal was so profitable, the dividends so luscious, and the market price of shares (when any were offered) so high, that the original capital of £130,000 in £200 shares was twice divided, and yet in 1831 each fraction fetched £690-695. Railway competition brought the price down, but £450 was the valuation accepted in 1846 by the newly formed North Staffordshire Railway, whose representatives were eager to forestall opposition from such a powerful corporation and to make full use of the canal during construction of their railway. The original capital, presently converted into £1,170,000 5 per cent railway preference shares, had been multiplied nine times.[11] Certainly the Thames & Severn was not in the position of the Trent & Mersey, but there was a chance of promoting conversion of the canal into a railway and of William Salt thereby realising a handsome profit on his very modest outlay.

At that date the Great Western was very far from being the magnificent railway it became in the first quarter of the twentieth century. With its associated companies, it had a virtual monopoly of the country west and south-west of London, but as most of its routes were indirect and its broad gauge isolated it from other railways, it was peculiarly open to the threat (blackmail is scarcely too strong a word) of competing lines.[12] This isolation engendered a corporate aloofness, and as the Great Western (and its public) was justifiably proud of the broad gauge and the speeds practicable upon it, the standard gauge companies were estranged. The two principally concerned were the London & North Western, maintaining the London & Birmingham's traditional championship of the Stephenson gauge, and the Midland, antagonised by the notorious inconvenience of the break of gauge in the west of England. Both these companies repeatedly sought to penetrate GWR territory with tracks of their own width.

It was in these muddy waters that William Salt proposed to fish, probably at the instigation of John Taunton, who had no belief in the future of the company unless the canal should be converted into a railway,[13] and whose previous experience qualified him to assess the possibilities, for he had surveyed a number of railway routes, had worked with Brunel, and had served as a district engineer in charge of construction of part of the Oxford, Worcester & Wolverhampton Railway.[14] But before taking such a serious step, as promotion of a competing railway, it was essential to consult someone even better qualified, and in 1861 the committee asked the advice of W. H. Barlow, previously resident engineer and by then consulting engineer to the Midland Railway. Barlow's opinion was that, in spite of the weakness of the Great Western's position in the Stroud valley, the prospects of a competing line did not justify the cost. Then, at the end of 1863, William Salt died, leaving, inter alia, 881 Thames & Severn Canal shares with prospects, and the Staffordshire collection valued at £30,000.[15] His widow, Helen Salt, retained the former but sold the latter which was acquired for the county and housed in Stafford Old Bank, thenceforward the William Salt Library.

Taunton kept the idea of a railway alive by a modest suggestion that the committee should lay "a line of Railway alongside the Canal to assist the traffic in times of drought". But the Salt family were fully aware of the possibilities and realised that the prospects were improving. The isolation of the broad gauge was nowhere more apparent than in South Wales, where many of the railways serving the mining areas were standard gauge, and whence, to avoid transhipment, coal trains ran to London by a roundabout route through Hereford, Worcester and Oxford. Both the LNWR and the Midland had many times tried, with limited success, to reach South Wales, and had also looked enviously at the Forest of Dean, but in 1863 these two companies and the Great Western concluded agreements (scheduled to an Act of Parliament) undertaking not to promote or assist "unnecessary" lines within each other's territory.[16]

What is or is not 'unnecessary' depends, however, on the point of view. The interpretation of the Midland, a wealthy and very progressive company, was wide, and certainly did not conform to the spirit of the agreements, scarcely even to the letter. While an offshoot from the Midland at Stonehouse to serve the factories and settlements in the Stroud valley (as the GWR did not) was locally considered necessary, the same could hardly be said of Midland attempts to reach out by direct or indirect means eastward and south-eastward through the Cotswolds.[17]

Then came the years 1864-1866, years of feverish railway speculation when many of the lines promoted had little justification, as they were designed to threaten existing monopolies rather than serve local needs. The broad gauge was especially vulnerable to such threats, some of which had so little substance that, as Richard Potter, then chairman of the GWR, said in words he would have done well to remember later, they were "made for sale, and for sale only".[18] Amongst schemes with some justification were those for an entirely new railway from London to Oxford, another across the Cotswolds to the Midland at Cheltenham, and a third westward from the Midland at Stonehouse, a chain which would have carried the standard gauge from London into the Forest of Dean and South Wales. It was the authorisation of two of the links, the Severn junction from Stonehouse with its proposed bridge across the Severn, and the East Gloucestershire from Cheltenham to Witney and a junction with the LNWR near Oxford, which led the Thames & Severn committee to re-examine the position in August 1865. Once again Taunton conferred with Barlow, and this time their diagnosis was favourable, for a Thames & Severn Railway from Stonehouse to the East Gloucestershire at Fairford would avoid the detour through Gloucester and Cheltenham and form an important link in "a through narrow gauge communication between Wales and the Metropolis" for coal trains.[19]

Between October and January engineers and promoters moved quickly, planning a railway nearly 24 miles long from Stonehouse, following the line of the canal between Stroud and Cerney Wick and thence bearing away to Fairford, with short branches to the canal wharves at Cirencester and Latton. The tunnel was suitable for a single track except at the western end, where a new piece, 154 yards long, would have to be driven on a curve to avoid the sharp bend by which the canal approached the old portal. East of the tunnel there were to be six miles of double track. Nowhere would the gradient be worse than 1 in 120 half that of the GWR. Between Stroud and Brimscombe, and Latton and Lechlade, the canal was to be maintained, but elsewhere the railway was to be laid as closely along its course as curvature permitted. Less for the company's own benefit than to forestall objection by the Stroudwater and Gloucester & Berkeley companies that closure of the summit would deprive them of a valuable supply of water, there was to be a conduit from the Thames Head pumping station to Baker's Mill reservoir and Brimscombe. The plan was to be financed by £375,000 of new capital and £125,000 borrowed on mortgage if required.

Before the Bill was introduced, Barlow and Taunton called upon Daniel Gooch, who had succeeded Richard Potter as chairman less than three months before, to discuss whether the Great Western would adopt the canal company's scheme. Gooch asked two questions: how much his company would have to guarantee "as an annual rental for the Canal as it is", and how much "if the Canal Company completed the Railway"? After deliberation, the Thames & Severn committee named £7,500 a year — a sum they had not earned since 1846 — as the price of blackmail, and £27,500 as the rental of their railway if used by the GWR, an amount calculated to provide 5 per cent on the proposed new capital and 1½ per cent on the old.

They had grossly over-estimated the strength of their position. How Gooch replied is not recorded but may be surmised, for in February the Bill was introduced, two of the Salt family providing the deposit money. During the parliamentary proceedings, representatives of the GWR showed great interest in the possible sources of the new capital, and one of them, cross-examining Taunton, pressed him hard upon this point: was there "some kind contractor —we will not say contractor — but there is some one who is looking with kindly eye upon you — some financial gentleman? . . . some one from whom you have some hope?" Only after repeated questioning was Taunton moved to admit: "There may be, but I might crush those hopes if I told you who it was". No such person casts the palest shadow on the proceedings, but speculative contractors were prominent in railway promotion at that date, and the possibility that there was one cannot therefore be ruled out. Quite evident, however, is the shadow of the Midland Railway: though the LNWR and the GWR engaged counsel, the Midland did not, and no officers of theirs appeared, as did William Cawkwell and James Grierson, general managers of the LNWR and GWR, to testify to the beneficial effects of the 1863 agreements.

The Bill sought unusually extensive running powers, covering some forty to fifty miles of six companies' lines, and as Barlow, questioned on these before the Commons committee, replied that their purpose was to enable the Thames & Severn Railway to work its own coal trains from the Forest of Dean to Oxford "without break of gauge, or any change of engine, or any stop", it became perfectly clear that the canal company's real intention was the revival of their lost trade in Forest coal. Both Grierson and Cawkwell told them there was no prospect whatever of their capturing the Oxford market with coal inferior to, and more expensive than, that obtained from Lancashire and the midlands — and their own records of the water-borne trade should have made it unnecessary for anyone to tell them. Grierson went further, adding that if they were to supply the entire Oxford market, the traffic would yield no more than £7,000 a year. The prospects were such as to appeal to a wealthy railway bent on extending its sphere of influence rather than to a financial speculator.

The Commons reduced the running powers, but although third reading was postponed because of opposition to the proposed diversion of water from the Thames Head springs to the valley of the Severn',[20] always a subject of controversy, they subsequently passed the Bill.[21] The Lords, however, made short work of it, their committee reporting on 19 July that it was "not expedient to proceed further".[22]

So ended the first serious attempt to turn the canal into a railway. Without doubt, it was a thoroughly unsound product of a period of feverish railway rivalry and speculation.

When the accounts had been paid, the coffers were empty. Liquidation of the contingency fund realised £7,020, more than £4,500 of which went to solicitors and parliamentary agents, and over £2,300 to engineers and surveyors — for a time Taunton had as many as three assistants.

Meantime the proprietors had allowed the canal to fall into a terrible state. The Commons committee had been told that a 40-ton barge was unable to carry more than 18 tons, that the summit became almost dry in summer, that for a period of fourteen weeks no traffic whatsoever had been able to pass. But with their resources drained, the committee were loath to authorise any expenditure. Discontent spread among the staff. Lock-keepers and labourers with little enough to do asked for — and got — a rise of 1s. a week, but seem to have been officious and discourteous.

In the summer of 1868, three men on their way from Manchester to London by water traversed the canal in a canoe and a skiff, and were the first of several pleasure-seekers to record their impressions.[23] Until reaching Saul, they had met with civility, and, as they used no locks, no canal charged them toll, but their reception on the Stroudwater and Thames & Severn was very different.

At Saul, the lock-keeper rudely demanded a pass, but as they had none and he could not issue one, he let them proceed. Higher up, another Stroudwater employee, having watched them carry their boats round his locks, claimed that they ought to pay "for passing over the company's water". Sharp words were exchanged, but appeased by a tip, he also let them go, shouting after them that they would "fare considerably worse" further on. This they found to be true. The Thames & Severn in adversity no more encouraged tourists than it had in Shelley's time, half a century earlier. At Walbridge, "a short, good-humoured man, with thin lips, extremely sharp, and inflexible to the last degree", claimed 10s. for the canoe, 30s. for the skiff because it had paddles as well as sculls, and £1 for the use of each lock. When they folded their hinged rowlocks, laid the sculls on the bottom, and assured him they would not use the locks, he accepted 10s. per boat and issued a ticket. Twice they were asked to show this, but each time so truculently that they made the official pursue them before doing so. As they ascended the valley, the close-set locks looking "over one another's shoulders" so wearied and discouraged them that at last they hired a cart to take them to Daneway, before plunging into the tunnel. Using paddles because the water was too shallow for sculls, they made their way for half-an-hour through darkness relieved only by the light of a single candle and the sight of the far portal gradually growing from the size of a pin point, until they emerged

Cerney Wick Lock and round house in 1947.

delightedly on limpid water "clear as crystal". Five-and-a-half miles further on they reached Siddington and met courtesy at last, a girl opening the locks and giving them "a ride down on the water". Discouragement of pleasure traffic was a pity, for, charmed by the Golden Valley, these three claimed the Thames & Severn Canal to be "one of the most beautiful [they] ever were on".

Nowhere did these men see any commercial traffic, although it was still passing, albeit in diminishing volume. The average annual tonnage for five-year periods between 1861 and 1875 was 58,317, 53,408, 45,543, but it is probable that not more than one-third of this went above Chalford.[24] Repeatedly Taunton tried to stir the committee's lethargy with proposals to increase it. One was for a siding from the Great Western to Brimscombe Port, so that coal could be obtained without passing over the Stroudwater, who were being awkward over charges. Another was for transferring railway wagons from the Midland to pontoons at Stonehouse wharf, whence they were to be floated to wharves and mills as far as Brimscombe, but the Stroudwater turned this down flat. There were suggestions for joint traffic arrangements with the GWR or the Midland which likewise got nowhere. But when outright sale was suggested to James Allport, the Midland general manager, he replied that although he could not consider that, he had extension to Stroud in mind and would therefore be interested in buying some of the company's land. When the committee decided they had no power to sell, Taunton himself intervened and bought "a large property" over which the branch was subsequently built.[25]

Nor were Taunton's proposals confined to co-operation with the railways. For two-and-a-half years he urged his committee to become carriers, but by then the canal was running at a loss and, despite his persistence, they remained unmoved. Even when he told them that Lechlade was being supplied with coal from Oxford, that the tunnel was in need of repair, that the summit ought to be relined, that the Thames Head engine had broken down and £60 would have to be spent "even to make it saleable", he could not stir them.

Taunton could do no more with that committee, but was there no prospect of invigorating it? He had made a reputation as a consulting engineer, and his advice on waterworks had been sought by the local authorities of Stroud, Cheltenham and Cirencester.[26] These activities, and keen membership of the Cotteswold Naturalists' Field Club, had brought him in contact with influential, public-spirited local people. It can scarcely have been anyone but he who drew together a few such men in support of his belief that canal trade could be increased by the company themselves becoming carriers. It was probably not difficult to interest them: by the seventies, any independent canal was valued as a means of restraining railway rates; a substantial interest in the Thames & Severn Canal could be acquired quite cheaply; effective improvement was bound to increase the value of the shares. An appeal to public spirit, coupled with a harmless speculation, was just the thing to attract landed and professional gentry, and it was they who provided all but one of the members of a new committee formed in April 1875.

They were Allen Bathurst, MP for Cirencester (later sixth Earl), Robert Anderson, Lord Bathurst's agent, Robert Ellett, a Cirencester solicitor actively engaged in the management of the upper Thames and subsequently President of the Law Society, John Dorington of Bisley, already embarked on his long and distinguished career in local government, and — of all men — Richard Potter, timber merchant of Gloucester and former chairman of the Great Western Railway.[27] It was a strong team to work with the enterprising engineer-manager, but it may be doubted whether more than one of them had an inkling where their policy would lead. Potter was a man of strong views who did nothing by halves: as chairman of the GWR he had "taken his position very seriously", and had resigned because the immense amount of work interfered with his own business;[28] quitting the railway board altogether two years later, he was by now bent on "emancipating this (our) district from a severe railway monopoly".[29]

The new committee soon adopted Taunton's policy, and in June 1875 began carrying, using horses, three boats, and the company's repair punt which could hold 18-20 tons. Unfortunately only scant details are available for the interesting period which followed as the relevant minute books are missing from the file, but incidental references show that the committee bought some steam barges the following year, that they owned a steam tug *Forward*, and that by 1879 they had a fleet of nine vessels, financed by the issue of £1,500 in debentures.[30] The registrations of these nine in 1879 suggest that *Spanker* was a second tug and that three were steam barges (*Frederick William*, *Excelsior*, *Edith*). Taunton had bought one of the tugs in London early in 1877, and the comment that he had great difficulty taking her up the Thames[31] was not encouraging.

In fact, the results were disappointing: the separate trade account showed a loss for the first two years, and the profits made in the four years thereafter were too small to offset the loss in maintaining the canal, so no interest was paid on the loan. It was not surprising: an inspection made in October 1875[32] showed that no boat drawing more than 2½ feet could have passed the summit, and it was common knowledge that in summer narrow boats had to reduce their lading to 18 tons, sometimes even 5 tons, to pass beyond Brimscombe. Moreover, the outlets eastward were in no better shape: the Wilts & Berks Canal was in such low water — in every sense — that no reliance whatever could be placed upon it; and as for the upper Thames, even rowing boats were liable to run aground in summer,[33] and trial voyages made in the winter of 1877-8 between Gloucester and Oxford with cargoes of timber (probably from Price, Potter, Walker & Co.'s yard) had resulted in boats lying aground near Shifford for more than two days until freed only with the help of "blocks, ropes, four donkeys and twenty men, and breaking £4 worth of ropes".[34] There was therefore no possibility of making the Thames & Severn Canal part of a through route in competition with the railway unless the river navigation should be improved; but there was little point in the Thames Conservators improving the river unless the canal also was to be improved, so no progress could be made without an assurance that the considerable sum of money needed for the latter would be forthcoming. The resulting stagnation led to the canal management being attacked in the press[35] by W. B. Clegram, engineer and superintendent of the Sharpness New Docks & Gloucester & Birmingham Navigation Company, a powerful combine formed in 1874 to acquire the new Sharpness docks, the Gloucester & Berkeley Canal and the Worcester & Birmingham Canal. Taunton did his best to parry the attacks while plans matured, but Clegram, who had served the Berkeley and Sharpness companies for forty-five years, was a fervent advocate of inland navigation[36] and, having established himself as the great panjandrum of the western waterways, was a persistent and formidable opponent.

The only possible source of money for the restoration of the canal was Richard Potter, who, in pursuit of his aim, was "prepared to give time, responsibility & capital" to making it an effective channel of communication. First, however, he had to secure a controlling interest. Even in 1875, he had been by far the largest of the new shareholders, holding 528, but finding that he could not acquire enough shares at the current price, he let it be known through Taunton that he was ready to pay "*something beyond* the present market values". The offer met with a very limited response, the other shareholders probably concluding that what was good for Potter would be good also for them. Helen Salt, who had already halved her holding, sold another fifty, but she retained 363, and Potter had added no more than 73. So he tried again. First he sent out a letter calculated to intimidate the older shareholders by alluding to the deteriorating condition of the canal and asking them to subscribe 10s. a share lest "the most serious consequences" should ensue; and then eighteen days later followed up with a second letter offering £3 apiece for shares that were, as he put it, "for all practical purposes . . . dead valueless to their owners". These tactics brought him another 450.

They were enough: with the support of Taunton, Bathurst, Ellett, Anderson, Dorington, Mrs Salt, and his partner C . B. Walker, he proposed the formation of a new company with a capital of £30,000 to take over the canal, discharge the debenture debt, and pay off dissentient shareholders at £3 a share.

A Bill was introduced, but met with vigorous opposition from the Great Western Railway, the Stroudwater Navigation, and the Sharpness company all of whom feared (with good reason) that Potter's new company was less likely to revive the canal than once again promote its conversion into a

Richard Potter, one of 'the last of the railway princes'.

railway. The Sharpness company indeed attempted to insert in the Bill a clause prohibiting any such action,[37] and rather than accept this, Potter dropped his proposal to form a new company and substituted a simple money Bill asking authority to borrow £30,000 on mortgage. Thereupon Taunton wrote to the opposing parties, countering what he called their "misrepresentations" with a disingenuous assurance that the Bill sought "no power . . . to convert the Canal or any part of it into a Railway", and asking them to withdraw their opposition. He was so far successful that the Bill was enacted in July 1879.[38]

The new issue of debentures raised £2,830 , and a genuine attempt was made to improve the navigation. But when nearly £1,000 had been spent upon the tunnel, there remained only a fraction of what was needed to restore the canal to some sort of working order, let alone to maintain that minimum depth of five feet which was what Clegram argued was essential. Clegram and Walter Stanton, one of his directors, foolishly chose to belittle the geological handicap from which the canal suffered, the former ridiculing Whitworth's reference to 'physical defects', the latter scornfully remarking that 'he knew a great deal was made of the porous character of the soil, but there was no real difficulty in applying a remedy; it was only a question of expense'. Although Clegram admitted that expense would be considerable, his confident assertion that "a moderate through-rate of charge" would make "the outlay . . . a remunerative one"[39] showed a lamentable ignorance, not only of the engineering difficulties, but also of the relatively small contribution through traffic had ever made to Thames & Severn revenue.

Potter retorted that without improvement of the River Thames or the Wilts & Berks Canal (of which there was no prospect), money spent on the Thames & Severn would be simply thrown away. He went further, expressing the opinion that whatever should be done to promote traffic by water between west and east, the GWR would reduce their rates for the commodity concerned with the result that "the reconstructed waterway . . . would not pay the cost of its maintenance".[40] As he knew the determination of railway policy from the inside, there was no reason to doubt the accuracy of his opinion: many people who then disbelieved him were to learn to their cost that Potter's attitude to canal restoration was a great deal more realistic than Clegram's.

Realising that he could never make the canal compete effectively with the Great Western, Potter determined again to promote its conversion into a railway. At this stage, some of his supporters withdrew. Mrs Salt, no doubt concluding that one such attempt was enough for her, sold out altogether: she would have done better to preserve her faith in Potter a little longer. Dorington, perhaps having developed a sentimental regard for the canal, perhaps entertaining scruples about Potter's intention, also sold out. Bathurst, Ellett and Anderson reduced their holdings. Potter became possessed of 1,642 shares with another fifty each in the names of his daughters, giving him undisputed control.

Potter was a much more formidable promoter of a Thames & Severn Railway than the Salts of 1866. His daughter, the social reformer Beatrice Webb, described him as loyal and transparently genuine but shrewd and impulsive with "a taste for adventurous enterprise and a talent for industrial diplomacy".[41] In what she described as "a maze of capitalist undertakings of which he was director or promoter", the railway industry predominated — indeed contemporaries called him one of "the last of the railway princes"; for at one time or another, he was a director of railways at home and in Holland and of a firm manufacturing signals, president of the Grand Trunk Railway of Canada, and chairman of the Gloucester Wagon Works. Among his friends were the great contractor Thomas Brassey and W. P. Price, former chairman of the Midland (and one of the partners in the timber firm).[42]

There was less excuse for a Thames & Severn Railway in 1882 than in 1866, as the Great Western had meantime converted its tracks in Gloucestershire and South Wales from broad to standard gauge and had begun work on the direct route to Wales through the Severn tunnel. Yet the prospects were more favourable, not only because of the personality of the promoter, but also because the new scheme was much more soundly based. On the eastern side, a new railway, later known as the Midland & South Western Junction and closely allied with the Midland, had been authorised and partly built along a line from Cheltenham through Cirencester to Andover. On the western, there was the Midland's projected branch to Stroud and their junction with the Severn & Wye Railway via the new Severn bridge. A Thames & Severn Railway was therefore assured of access to Forest collieries at one end and to markets around Cirencester at the other. An important adjunct to the scheme, however, was the concurrent promotion of the South Wales & Severn Bridge Railway, which afforded the glittering prospect of the Thames & Severn Railway being used by trains on their way from South Wales to the London & South Western, carrying coal destined for London or the bunkers of steamships in the busy port of Southampton.[43]

Potter's Bill of 1882 proposed to close the whole canal and to build "in part upon the site" a railway 14¾ miles long from Stroud to a junction with the Cheltenham-Andover line at Siddington, financing it by the issue of £180,000 in shares with power to borrow £60,000 on mortgage if required. Running powers were sought from Stonehouse to Stroud and Siddington to Cirencester, but more to the point was a clause authorising arrangements for the Midland to manage and work the line. Alliance with the Midland, no doubt greatly helped by Potter's lifelong friendship with Price, was indeed already close, for Potter had consented to do what the old canal committee had refused to do — sell part of Stroud wharf for the site of the Midland station, which moreover his Thames & Severn Railway was to share. For his undertaking to do this and promote his railway Bill, Potter had received a personal gift of £2,000 from the Midland company.[44]

The Bill was introduced in the House of Lords in February 1882,[45] but as it turned out, its fate depended not on the attitude of the Midland Railway, but on that of the Sharpness New Docks & Gloucester & Birmingham Navigation Company. The Sharpness company had become involved in railway enterprise by the action of their forerunner, the Gloucester & Berkeley, in subscribing to the cost of the Severn bridge, which they expected to increase the revenue of their new docks by providing the railway facilities hitherto lacking at the quaysides and by bringing ample supplies of coal from the Forest of Dean. As trade in Forest coal alone was unlikely to make the costly bridge remunerative, the Sharpness board were unanimously in favour of the projected South Wales & Severn Bridge Railway. But whereas the chairman, W. C. Lucy, and five other directors were adherents of Potter's Thames & Severn Railway, there was a group led by Stanton which shared Clegram's view that water communication with the Thames ought to be preserved, and who therefore pinned their faith on plans which they knew were about to be brought forward for a different railway running eastward from the bridge which would leave the Thames & Severn Canal unimpaired.[46] So when the Sharpness board met to determine their attitude to Potter's Bill, there was a close division: Lucy's party was defeated by seven votes to six, and the board adopted a policy of "the most decided opposition".[47]

Potter, who was a large shareholder in the Sharpness company, strove hard to get this decision reversed, promising, if the board did so, to increase his subscription to the South Wales line and use his considerable influence in raising the sum needed to meet its parliamentary expenses.[48] Furthermore, backing persuasion by threat, he roundly declared that if the Sharpness board would not withdraw their opposition and his Bill should be thrown out, he would "run the Canal dry";[49] it was nearly that already, and there seemed little to prevent him, by then almost sole proprietor, from doing as he said.

Nevertheless, after a series of divisions in board meetings and a specially convened assembly of shareholders, the Sharpness company maintained their attitude.[50] Lucy vacated the chair and was succeeded by Stanton, who proceeded to organise opposition to the Thames & Severn Railway Bill with such vigour that a petition was submitted not only on behalf of the Sharpness company, but also of the Stroudwater Navigation, the Wilts & Berks Canal, the Severn Commissioners, the Staffordshire & Worcestershire Canal, and the Birmingham Canal Navigations,[51] a coalition which, with variations in its composition, survived as the Allied Navigations for thirteen years or more and played the leading role in the farce that was staged.

This was bad enough, but far more formidable opposition then suddenly developed. The Allied Navigations enlisted the support of other canal authorities and approached the Board of Trade, with the result that a deputation representing the Canal Association was received by the President himself, Joseph Chamberlain. They told Chamberlain that in their opinion Potter's proposal violated the Regulation of Railways Act of 1873, a statute intended to prevent any more canals being converted into railways or acquired by railway companies without express authority. To counter this move, Potter also obtained a personal interview with Chamberlain, and pointed out that a primary reason for the Bill was that it was impossible to maintain the canal in operation except at a loss of about £550 a year. Recognising Potter's problem but not accepting his method of resolving it, Chamberlain then asked the Allied Navigations whether they had any definite proposition for keeping the Thames & Severn Canal open.[52]

The only members of the alliance in a sound financial position were the Staffordshire & Worcestershire and Stroudwater companies, for even the Sharpness, saddled with their unproductive investment in the Severn bridge and unremunerative operation of the Worcester & Birmingham Canal, had made no distribution on ordinary shares and had paid preference dividends from reserves.[53] The Birmingham Canal Navigations had been heavily subsidised by the LNWR for six years, the Severn Commission by the GWR for more than twenty-five. The Wilts & Berks, sold for a song, had recently been leased by a group more sanguine than perspicacious who soon found it could not pay its way.[54] Nevertheless representatives of these Allied Navigations met, under Clegram's presidency, on 27 February 1882 and formulated proposals of surpassing impudence. They would manage and maintain the Thames & Severn Canal for twenty-one years, receive its revenues, pay interest on its debentures, distribute 80 per cent of any profits to its shareholders, and, as even they must surely have realised was far more likely, meet any loss by contributions graded according to the benefit each navigation received from through traffic.[55]

It was a remarkably naïve attempt to confiscate the property.[56] Potter had no intention of allowing that, nor had he any faith that the Allied Navigations possessed either the resources or the capacity to restore the Thames & Severn Canal and in this, subsequent events proved him correct. But Joseph Chamberlain was sufficiently impressed, and informed the Canal Association that he would oppose the Bill. Potter therefore had no option but to withdraw it,[57] and concluding that further negotiations with the Allied Navigations offered no prospect of advantage to anyone, certainly not to himself, he turned to Paddington, and offered his shares, now at least 2,030, at £8 apiece to the company whose monopoly he had striven so hard to break. He found Sir Daniel Gooch, who for the second time faced the danger of the Thames & Severn becoming a railway in close association with the hostile Midland, ready to negotiate. On 11 May 1882 the Great Western board agreed to purchase Potter's shares (and as many of the remaining 420 as the holders should care to part with) for £7 10s. each, the difference of 10s. a share being retained by the GWR to meet the debenture debt and cover any loss in working the canal. They also stipulated that Potter should hand over the £2,000 he had received from the Midland.[58]

He could well afford to do so: according to the secretary of the Sharpness company, Potter had acquired some of his shares "for as little as 5s." each;[59] for the majority he had paid £3; his outlay had been perhaps £4,500, so that he cleared about £10,000.[60] He was the first, and last, to make a considerable profit from Thames & Severn Canal shares. He was also the first and by far the most successful of those who, having discovered that the canal was nothing but a heavy liability, sought to pass the buck while there still remained other hands willing to accept it.

REFERENCES

1. MacDermot op cit, vol. 1, pp. 9-10, 162-3. *Salisbury & Winchester Journal* 16 November 1835, advertisement.
2. Clapham op cit, p. 398. Report of the Royal Commission on Canals & Waterways 1906-9 vol. 8, p. 9, gives a lower proportion of about one third.
3. Some of the advantages of railways and disadvantages of canals were discerned by Sutcliffe (op cit, p. 73) as early as 1816, by Dupin in 1820 (op cit vol. 1, pp. 220, 229-230), and by the author, himself a shareholder in the Kennet & Avon Canal, of *Observations on the general comparative merits of Inland Communication by Navigations or Rail-Roads*, 1825. They have been examined by Pratt, E. A., *British Canals*, 1906; by Jackman op cit, vol. 2, 1916, pp. 489-497; and by Clapham op cit, 1926, pp. 82-4, 396-9.
4. The point that "mere boatmen owning the boat . . . of course . . . could not take much responsibility" was stated by W. B. Clegram, then engineer and superintendent of the Sharpness New Docks & Gloucester & Birmingham Navigation Co. before the Select Committee on Canals 1883 (Report p. 102).
5. In my opinion, insufficient attention has been paid to this vitally important matter, although figures for stoppages between 1 June 1905 and 31 May 1906 were reported to the Board of Trade by various canals and were quoted in the Report of the Royal Commission on Canals & Waterways 1906-9 (vol. 1 part 2, Appendix 1, statement 4), and stoppage for repair was the subject of discussion in vol. 3, p. 182, and vol. 5 part 2, p. 39.
6. Abstracted from T & S Collection: Siddington Returns 1809, 1822, 1829, 1839; Weekly Return of Arrivals & Departures of Vessels at Brimscombe Port 1827, 1830; Letters from J. Denyer to J. S. Salt 1828.
7. MacDermot op cit, vol. 1, pp. 162-174. Reports of the General Meetings of the Cheltenham & Great Western Union Railway 1836-1843 and printed letters of 1836 addressed to shareholders from the directors and from a supporter of the rival route are in GCL. *Railway Magazine* 1950 vol. 96, pp. 79-82, article on Sapperton Tunnel by H. G. W. Household.
8. MacDermot op cit, vol. 1, pp. 567, 574. Forest of Dean Railway General Meeting Book 8 January 1852.
9. Aris's *Birmingham Gazette*, 25 August 1845.
10. DNB William Salt, 1805-1863.
11. Trent & Mersey Canal Acts 6 Geo. III cap 96, 42 Geo. III cap 25, 7 & 8 Geo. IV cap 81. North Staffordshire Railway Report of meeting of proprietors, 23 September 1846 (BTR). T & S Collection, papers recording prices of various canal shares. Pratt op cit, pp. 26-27.
12. See MacDermot op cit Vol. 2, p. 24.

13. See Report from the Select Committee on Canals 1883, p. 118, evidence of J. H. Taunton.
14. *Proceedings of the Institution of Civil Engineers* Vol. 112, p. 361.
15. DNB William Salt, 1805-1863.
16. MacDermot op cit vol. 1, pp. 436-7, 447-8, 528-530, 538-541, 551-2, 581.
17. Ibid vol. 2, pp. 24-7.
18. Ibid vol. 2, pp. 23-4.
19. The following paragraphs are based on: the draft of the Thames & Severn Railway Bill 1866, the printed minutes of the Proceedings of the Commons Committee in April 1866 (from which the quotations have been taken), and the Bill as amended in committee, which are in the T & S Collection; T & S Committee Register, 22 January 1866 which reported the interview between Barlow, Taunton and Gooch which had taken place that morning; and articles in GJ, 31 March, 14 April, 5, 6 May, reporting on the local railway schemes and their fate.
20. GJ, 19 May 1866.
21. JHC vol. 121, pp. 321, 326, and entries between pp. 47 & 315.
22. JHL vol. 98, p. 593, and entries between pp. 22 & 555.
23. *The Waterway to London, as explored in the "Wanderer" and "Ranger", with sail, paddle, and oar, in a voyage on the Mersey, Perry, Severn, and Thames, and several canals*, 1869, pp. 61-73.
24. This was so in 1885-7, see pp. 1888 Accounts & Papers Vol. 89, p. 576.
25. *Proceedings of the Institution of Civil Engineers* vol. 112, p. 362.
26. Ibid pp. 361-3.
27. Gardom, E. T., *A Brief History of the Thames & Severn Canal*, 1901 (published by the Gloucestershire County Council), p. 5. GJ 8 April 1911, obituary of Sir John Dorington. For other information, I am indebted to the Secretary of the Law Society and the Librarian of the Bingham Public Library, Cirencester.
28. MacDermot op cit vol. 2, pp. 30-32.
29. Circular letter from R. Potter to the Shareholders of the Sharpness New Docks & Gloucester & Birmingham Navigation Co. 14 March 1882 (GCL).
30. Thames & Severn Canal half-yearly statement to 4 April 1882 (BTR). In GCRO: General Trade Account January 1876; Certificates of Registration of Canal Boats 1879; Brief to Promote the Company's Bill of 1879; Report to the Allied Navigations re. Thames & Severn Canal 1894, p. 27.
31. *Gloucestershire Chronicle*, 10 February 1877.
32. Report to the Allied Navigations re. Thames & Severn Canal 1894, p. 21, quoting from a report made by E. Leader Williams, M. Inst. C.E., 7 October 1875.

33. Gardom op cit, p. 4.
34. *Gloucestershire Chronicle,* 10 February 1877. Reports on the voyages are in the T & S Collection.
35. See *Gloucestershire Chronicle* 17, 24 February, 3, 10 March, 1877.
36. *Proceedings of the Institution of Civil Engineers* vol. 98, p. 389.
37. See pp. 1886 Accounts & Papers vol. 60, p. 694.
38. Act 42 & 43 Vict. cap 71.
39. See *Gloucestershire Chronicle* 10, 17 February, 3 March 1877.
40. Circular letter from R. Potter to the Shareholders of the SND & G & B Co., 14 March 1882 (GCL). Although these words were written by Potter in 1882, they undoubtedly represent conclusions to which he had been driven several years earlier.
41. Webb, B., *My Apprenticeship*, 1926, pp. 2-10, 20, 342.
42. GJ 9 January 1892, obituary of R. Potter.
43. Circular letters from R. Potter (14 March 1882) and W. C. Lucy (30 March 1882) to the Shareholders of the SND & G & B Co. (GCL).
44. PP 1886 Accounts & Papers vol. 60, p. 694. GWR Minute Book of the Directors 11 May 1882.
45. JHL vol. 114, pp. 23, 27, 31. See also JHC vol. 137, pp. 34, 86.
46. Circular letters from W. G. Postans to the Shareholders of the SND & G & B Co. 23, 30 March 1882 (GCL). See also *Railway Magazine* vol. 102, p. 668, article on the London & South Wales Railway by B. G. Wilson.
47. SND & G & B Minute Book, 1 February, 5, 13, 19 March, 19 April 1882. See also letter from W. C. Lucy to the Shareholders 30 March 1882.
48. Circular letter from R. Potter to the Shareholders of the SND & G & B Co. 14 March 1882.
49. Circular letter from W. G. Postans to the Shareholders of the SND & G & B Co. 30 March 1882, quoting Potter's remark (GCL).
50. SND & G & B Minute Book 1 February, 29 March, 5, 13, 19 April 1882.
51. A copy of the petition is among BTR.
52. SND & G & B Minute Book 1 March 1882. pp. 1886 Accounts & Papers vol. 60, pp. 695, 698; *Stroud Journal*, 18 February 1882.
53. See circular letter from W. C. Lucy to the Shareholders of the SND & G & B Co., 30 March 1882.
54. See Pratt, *British Canals*, p. 60; Hadfield, *British Canals*, pp. 215-7. Hadfield, *Canals of Southern England*, p. 302.
55. PP 1886 Account & Papers vol. 60, pp. 698-9. See also SND & G & B Minute Book 1 March 1882.
56. Taunton twice referred to the proposals of the Allied Navigations as "confiscation" when giving evidence before the Select Committee on Canals 1883 (see Report pp. 115, 118).

57. See circular letter from R. Potter to the Shareholders of the SND & G & B Co., 14 March 1882. The proceedings in the House of Lords ended with withdrawal of the Bill on 2 May 1882 (JHL vol. 114, pp. 55, 66, 71, 73-4, 88, 130).
58. GWR Minute Book of the Directors 27 April, 11 May 1882.
59. PP 1887 Accounts & Papers vol. 72, p. 897.
60. I think it probable that Potter's first 528 shares were acquired for 5s. each, costing him £132. For his next 73 he may have paid around 7s. each, i.e. £25 11s. For the remaining 1,429 he almost certainly paid £3 each, i.e. £4,287, making a total of £4,444 11s. The GWR would have paid him £15,225.

CHAPTER THIRTEEN

THE DEATH AGONIES
I: MISPLACED TRUST

What was the legal position after Sir Daniel Gooch had bought Potter's shares? Undoubtedly he had contravened one intention of the Regulation of Railways Act of 1873, but he had done so only to prevent conversion of the canal into a railway, thereby complying with another of the intentions of the Act. Evidently there were doubts, for care was taken to preserve the corporate existence of the canal company and to transfer the shares so discreetly that the railway company's title was not immediately apparent.

By no means all the old shareholders seized the opportunity to dispose of their unprofitable shares profitably, even Potter, Earl Bathurst, and Taunton retaining a few; but 2,205 were acquired by the Great Western, and no less than 2,149 of these were held jointly in the names of Sir Charles Mills and the Hon. P. C. Glyn, partners in the banking house of Glyn, Mills & Co. The remainder were transferred to various GWR shareholders for example, ten to Charles Mortimer and three to Thomas Holland, both of whom became members of a small sub-committee of three responsible for the management of the canal. Holland claimed a special interest in inland navigation and represented the GWR on the River Severn Commission; and as neither he nor Mortimer was active, in railway management at that date, it was not at all obvious that as members of the Thames & Severn committee they were Great Western nominees.[1]

Certainly the Allied Navigations were puzzled and irritated by the sudden failure of their attempt to gain control of the Thames & Severn Canal; certainly the Sharpness board suspected that the GWR was involved, but they were unable to confirm their suspicion.[2] As late as June 1883, Clegram, giving evidence before the Select Committee on Canals; answered the question "And there is no control over [the Thames & Severn] by a railway?" with the assurance "There is not", and the further question "The Thames and Severn Canal Company are entirely free from railway control, are they not?" with the words "Quite so".[3] The Sharpness company's inability to establish exactly how they had been outwitted more then twelve months before is astonishing, but it goes far to explain their relentless harrying of the Great Western when at last they learnt the facts.

In the meantime the Sharpness board, backed no doubt by the Stroudwater company who stood to gain more than any of the other navigations from an increase of traffic along the Thames & Severn but who nevertheless played a less obvious role throughout, were determined to pursue the aims of the Allied Navigations a stage further and resolved on very high-handed action indeed — no less than compulsory transfer of the management of the Thames & Severn into the hands of a joint committee on the ground that the existing management was not maintaining it in an efficient state. They announced their intention of applying for parliamentary authority to do this, and at the same time their secretary, Henry Waddy, engaged in an acrimonious correspondence with John Taunton which the latter construed as "an attempt to interfere with the details of the management" in order to buttress their case.[4] In November 1882 the Allied Navigations even asked for a report on "the measures essential to secure the efficiency of the Canal" and commissioned Clegram, Henry Marten (engineer to the Severn Commissioners) and W. J. Snape (clerk to the Stroudwater Navigation) to draw it up for them. Both Clegram and Marten were members of the Institution of Civil Engineers. As relations with the management at Brimscombe were anything but cordial, the three were not in a position to inspect the canal. Marten, however, had done so some ten months previously, and so their report submitted early in 1883 was based on his knowledge and on the results of enquiries, mostly answered by traders who complained of silting, leakage, shortage of water, and inability to carry full cargoes.[5]

Fairly enough, the three men remarked that the canal was in the same "dilapidated condition" as it had been for some years, but in stating that it was lack of depth which had caused 60-ton barges to be superseded by narrow boats, they showed a deplorable ignorance of the facts, and, in Marten's case, an astonishing lack of observation, for he ought to have noticed that most of the locks east of Chalford had been shortened, precluding the use of Thames barges of that size, and he might have deduced that the work had been done because such vessels had ceased to use the line; as already related, this had been done forty years earlier when the canal was in its finest condition. Improbable as may seem, Clegram, Marten and Snape ended their report by stating casually, in a postscript, that they estimated "the total cost of putting the Canal into a fair condition of repair at £10,000". Yet this figure, reached without the most cursory examination of the works, let alone the meticulous care essential in forming estimates, remained accepted and unquestioned for a dozen years with most unhappy results.

Sometime in the late summer or autumn of 1883, the Allied Navigations (or Henry Waddy, who was in fact their active agent) dismounted from their high horses and reverted to negotiation, and it was about the same time that they became aware of what had really happened to the Thames & Severn

Canal. The Birmingham Canal Navigations had withdrawn from the alliance — probably under pressure from the LNWR who subsidised them — and had been replaced by the impecunious Wilts & Berks, but in the end only the Sharpness and Stroudwater companies were represented at a meeting with Holland and Taunton. The Allies' terms were stated, but as these now included a proposal to borrow £12,000 for restoration of the canal, and to make repayment a first charge on receipts, the terms were even less acceptable than before. Freely admitting his association with the GWR, Holland refused to discuss such confiscation, and proceeded to an unflattering analysis of the financial state of the navigations, pointing out the failure of the Sharpness to pay their preference shareholders in full and questioning their right to accept any additional liability meanwhile. Feelings ran high and the meeting broke up, having served only to exacerbate both sides.[6]

For three years thereafter there was a lull, during which the Great Western more than once stated their readiness to transfer the canal "on reasonable terms", bearing in mind the expense they had incurred to save it from closure,[7] and the committee of management struggled with the thankless task of maintaining the canal in operation executing general repairs, stopping extensive leaks, and renewing lock gates.[8] In fairness to the railway company — to whom little fairness was ever shown — they never, so Holland stated, "attempted to influence" his conduct of the canal management in any way, still less to divert traffic from water to rail, nor would he have allowed them to do so.[9] There was indeed little point in such action: they had not undertaken to maintain the canal revenue, and their interests were best served by letting the canal earn what revenue it could and cutting the losses as low as possible.

The latter Holland's management achieved with some success: for although in 4½ years to September 1886 there was an overall loss, the concluding twelve months actually showed a profit of £569.[10] As always, however, leakage of the summit was the great problem, absorbing more than 60 per cent of the water pumped at Thames Head, and after Taunton had stated that to reduce it effectively would cost £3,800,[11] the committee decided to cut expenses still further so that "all available means" might be applied to improvement.[12] This policy was carried to great lengths and had results the committee were unlikely to have foreseen. In the winter of 1885-1886, several lock-keepers were dismissed, the carrying trade was discontinued, one of the two enginemen was discharged and the other laid off for the season; and when the summer came, resumption of pumping was delayed on the ground that it was virtually useless as long as so much water ran to waste. Nor was this all, for in the latter part of 1885, the committee terminated Taunton's appointment as engineer and manager, and, although they asked him to state his own terms for re-engagement as engineer only, this proved to be a tactical error.

Only a short while earlier, Taunton had had a brush with the Great

Western accountants over the archaic methods of accountancy followed at Brimscombe, and another with the canal company's clerk on the awkward subject of undefined 'perquisites' or 'privileges'. He may therefore have felt that termination of his appointment was not only a poor return after thirty-three years' service, but also a reflection on his conduct of the business. Aged 64 and obviously embittered, he turned down the committee's suggestion, and thereafter (he lived until 31 January 1893) used his considerable local influence to oppose their policy.[13]

The Sharpness company eagerly seized on these economies as evidence of the malign influence of the railway. Aided now by Taunton, they stirred up local agitation to such effect that the Board of Trade was bombarded with letters and memorials claiming that the Great Western's acquisition of the canal shares was 'ultra vires', and urging the Board to support the proposals for transferring control to the Allied Navigations. Some of what was said was true, but much was misconstrued, and not a little was based on mere suspicion of intent, but for more than ten months the Board of Trade was obliged to pass accusations and explanations to and fro between the GWR secretary on the one hand, and Waddy and a gaggle of solicitors, memorialists and disgruntled traders on the other.[14]

Meantime there was a change of government. A Liberal President of the Board of Trade, A. J. Mundella, had hitherto handled the voluminous correspondence, but it was a Conservative, Lord Stanley, who received a deputation with yet another memorial on 29 November 1886. The meeting was an anti-climax, for the memorialists had already reached the conclusion that the railway company's action had not been illegal, and the deputation was told that the law officers of the Crown held the same opinion.[15] By this time, repairs had been completed, pumping resumed, a minimum depth of 3 feet achieved, and the Great Western's position was a good deal stronger — certainly strong enough for their secretary to suggest a formal Board of Trade enquiry.[16]

Apart from recommending the Allied Navigations to re-open negotiations with the railway, an idea they utterly refused to consider,[17] there was little else the Board could do but have the canal inspected. This was done by Courtenay Boyle, the Board's assistant secretary (later permanent secretary and a KCB), accompanied by Thomas Holland, and his report was submitted in October 1887. He "did not consider it proved" that the GWR had deliberately neglected the canal, for he had found the reaches below Brimscombe deep and well maintained. But the summit was narrowed by weeds and silt, and the depth considerably reduced after prolonged drought, and as any substantial improvement would be "a matter of extreme difficulty" involving "a large outlay" which the existing management would certainly not undertake, he believed that control should be in the hands of "persons more nearly interested" in maintaining it in good condition; he also therefore suggested that the Allied Navigations should re-open negotiations.[18]

The following year, the directors of the Stroudwater Navigation, acting alone, did indeed try a direct approach, making an offer of £3,000 which the Great Western understandably rejected as wholly inadequate.[19] But, as it was obviously useless expecting more of the Board of Trade, the agitation ceased for several years, until, indeed, the dying embers were fanned in September 1893 by another memorial signed "by 152 Traders, Merchants, Boat Owners, and others" — not many of whom are likely to have been actual users of the canal[20] — urging local MPs 'to do something about its unsatisfactory management and condition".[21]

It was the canal committee themselves, however, who blew the embers into a blaze by announcing on 28 December 1893 that *in two days time* the canal east of Bell Lock, Chalford, would be closed "until further notice".[22] Locally, it was assumed that 'further notice' was never likely to be given, and that the intention was to abandon all that part. As has already been pointed out, about 43 per cent of the trade was confined to the Golden Valley even in the busiest days of the navigation; the proportion had greatly increased since then, figures quoted by Courtenay Boyle showing that 67 per cent of the traffic and 54 per cent of the earnings derived from Chalford or below.[23] No commercial considerations justified the expense of maintaining the summit any longer.

In spite of the short notice, the opposition quickly mobilised their forces. Within three days, the Cirencester Local Board, the Stroud Chamber of Commerce, landowners and others from the countryside between Chalford and Inglesham, had all held meetings from which letters of protest or appeal had gone out to the canal company, the Board of Trade and the Allied Navigations.[24] It was as though accusing ghosts had risen to rouse the countryside in protest at the undoing of what they had planned and built and assiduously cared for. In fact, however, all the protesters were as certainly dominated by the fear of railway monopoly[25] as Richard Potter had been, but whereas his feet had been firmly on the ground, their heads were in the clouds.

The outcome of all this activity was that "a very large and influential Deputation" visited the President of the Board of Trade — A. J. Mundella once again — on 10 January 1894. Mundella listened sympathetically to at least nine speakers, one of whom, the MP for Stroud, made the pertinent suggestion that as the canal "had become . . . derelict", "if a Local Authority made an application . . ., the Board of Trade had power to transfer the undertaking to that Local Authority". He was supported by the MP for Cirencester, but although Sir John Dorington, then MP for Tewkesbury and first chairman of the recently formed Gloucestershire County Council, may have shown interest in the idea (which he was later to use), Mundella does not appear to have done so, certainly not the interest he showed in Waddy's statement that there was no doubt the Allied Navigations "would undertake to raise money to buy the Canal if the Great Western had no further need for it".[26] Apart from the Stroudwater company's private offer, the navigations had never before suggested 'buying' the canal, merely paying "a strictly nominal rent" for it,[27] and Waddy's statement therefore appeared to be, as Mundella remarked, a very important one. In fact, however, it was either a slip of the tongue or a deliberate misrepresentation.

Mundella closed the interview with sympathetic, woolly verbiage. "Nothing could be more disastrous", he remarked, than canals falling into the hands of the railways "when in the interests of trade they ought to have been kept open, developed and improved so as to maintain a fair and open competition" with the railways.[28] Did he know what the LNWR was paying to the Birmingham Canal Navigations, or the North Staffordshire to the Trent & Mersey, or the GWR to the Kennet & Avon?[29] Did he stop to think where the money could have been found to maintain the Thames & Severn in first class condition since 1874? Nevertheless he assured the deputation that the Board of Trade "would do anything they could to bring about the desired result" of opening "the waterway so that all the difficulties and losses which had been indicated might come to an end". How to achieve this was another matter: "It was quite clear that the Company ought to maintain the navigation or abandon it. He was not sure whether they could not be compelled to maintain" it. He concluded with a promise to find out the canal company's intention and to let the deputation know the result.[30]

In the meantime the canal shareholders had been called to meet at Reading on 16 January to consider whether the summit should be maintained or "certain portions of the Canal" should be closed. The *Wilts & Gloucestershire Standard*, which played a prominent part in the campaign against the railway company, made game of the announcement: "Why the 'proprietors' should be solemnly invited to journey from Paddington to Reading 'for the purpose of *considering*', when they have already so decisively *acted*, may be apparent to the railway mind, but plain people will be inclined to regard the affair as partaking of the character of a farce".[31]

It is unlikely that Holland and his colleagues were quite as naïve as the *Standard* assumed. For one thing, Clegram had bought 34 Thames & Severn shares in 1882 and these had been distributed among seven members of the Sharpness board[32] (this, in turn, led to a wider distribution of the Great Western's shares), so it was no longer safe to assume that a meeting of proprietors would adopt the committee's policy without demur. As for the meeting place, it is quite possible that Reading had been chosen out of consideration for those Sharpness shareholders, who would have resented being summoned to Paddington, just as they had resented the recent transference there of the clerk's office and the company's books. As for the timing of events, it is likely that only two days' notice of closure had been given so that there should be no opportunity for anyone to intervene and prevent it, and that the meeting of shareholders was called for a later date so that the strength and plans of the opposition should be known beforehand. In point of fact, the Great Western, having already ascertained that it was "not possible to make the canal a paying concern" and that whatever money should be spent "would be unremunerative",[33] had concluded that it was time for them to pass the buck. They were very wise.

The conduct of the shareholders' meeting was exemplary. Walter Robinson, a director of the GWR, pointed out that the railway company had never been "the owning company of the Canal" and could not therefore be liable for its maintenance. Thomas Holland, in an excellent address, explained that he had hitherto rejected the terms proposed by the Allied Navigations because he was unwilling to prejudice the shareholders' position before finding out whether or not his committee could maintain the canal as they had found it, but as they could not and the company had no money (Mortimer remarked that they did not even "know how to pay next week's wages"), he was now asking for authority to re-open negotiations. With the help of such "a strong body of gentlemen" as the allies, he said (probably with his tongue in his cheek), it should be possible not only to restore the canal but also to persuade the Thames Conservancy to make the river once again navigable by laden vessels between Lechlade and Oxford. Two resolutions were then carried, one authorising the committee to make such arrangements with the neighbouring navigations as they should think fit, and the other to strengthen their negotiating position by empowering them to close parts of the canal either temporarily or permanently. Hubert Waddy, Henry Waddy's nephew and successor as secretary to the Sharpness company, said he was sure the navigations would be willing to negotiate, and, in contrast to his uncle's high-handed ways, courteously moved a vote of thanks to Holland for the information he had given and his conduct in the chair.[34]

In this clement atmosphere, negotiations were resumed. In the summer, the Thames & Severn Canal was inspected from Inglesham to Brimscombe by W. J. Snape, W. J. Ainsworth (manager of the Wilts & Berks), G. W. Keeling M. Inst. C.E. (engineer and general manager of the Sharpness company) and F. A. Jones (another engineer in the latter company's service). They produced for the Allied Navigations a detailed report of great importance at the time and of considerable interest now for comparison with earlier records of the canal property. Perusal of it suggests that they did their work thoroughly and prepared their estimates with meticulous care: all the same, the figure of £10,309 for restoration of the canal to "fair working order" is strikingly similar to that put forward so casually by Clegram, Marten and Snape eleven years earlier. The four had taken into account all known points of weakness, but they remarked that there might well be others, and in particular that "doubtful points" were the allowances for eliminating leakage from the tunnel and summit.[35] Nevertheless, no one at that time appears to have had the least suspicion that this important reservation could invalidate the entire estimate.

In January 1895, agreement was reached. The GWR consented to transfer the shares on condition that no part of the canal property should ever be used for railway purposes, that the outstanding debenture debt (then £2,830) should be paid off at the rate of 15s. in the pound, and that after ten years there should be payments from net revenue to the former shareholders (thenceforward known as contingent annuitants) — there was no difficulty in reaching agreement on this point because "neither side anticipated . . . any hope of such payment".[36]

Five navigations (the Severn Commissioners and the Sharpness, Stroudwater, Staffordshire & Worcestershire and Wilts & Berks companies) were joined in the venture by six public bodies (the City of Gloucester, the Urban Districts of Stroud and Cirencester, and the County Councils of Gloucestershire, Wiltshire and Berkshire). Together they formed the Thames & Severn Canal Trust, incorporated by Act of 1895 with authority to borrow £15,000 to set the canal in order.[37] There were detailed arrangements for the distribution of revenue, most of which had no significance whatever, but it is interesting that during the negotiations the public bodies had managed to ensure that payment of rents, rates and taxes had priority even over maintenance and the wages and salaries

The Trial between the locks at Walbridge on her way from London to Eastington, 11 April 1899.

of the Trust's unfortunate employees. More significant were the arrangements for covering any deficit. The accounts of the Trust[38] show that in practice slight variations were made in these, the navigations covering the first £600 in agreed proportions (a quarter falling on the Stroudwater), and the public bodies meeting the next £600 (three-eighths provided by the Gloucestershire County Council), any further deficit up to £1,200 being likewise equally divided. Furthermore the Act provided that if income should fail to cover expenses in two successive years, the Trust could apply to the Board of Trade for an order to abandon the canal and sell the property.

As it was generally recognised that local traffic to and from points east of Chalford would be slight, hopes were pinned on the development of through traffic with Oxford. The success of the venture therefore depended on the condition of the River Thames navigation. Happily, the Thames Conservancy had set to work above Oxford even before formation of the Trust, completing a new lock at Radcot in 1892 and beginning another at Northmoor. A mutual understanding existed between Trust and Conservancy that each would improve its navigation for the benefit of both, and so, with more extensive river works in mind, Keeling, Jones, Hubert Waddy and the Conservancy's engineer set out from Lechlade for Oxford in April 1895, travelling ten miles in a rowing-boat before they could use a steam launch, discussing as they went plans for new locks, clearance of weeds, removal of shoals and sharp bends, and restoration of the towing-path. Between then and 1898, Northmoor Lock was completed and two others built at Grafton and Shifford, the Conservancy spending in all some £20,000 upon the upper district. Although four flash-weirs remained for some thirty years, the navigation between Inglesham and Oxford was better than it had ever been before.[39]

The Trust was administered by a committee of eight members, five appointed by the navigations and three by the public bodies. Their chairman was Sir William Marling, a prominent manufacturer in the Stroud valley, a director of the Stroudwater Navigation, and chairman of the Sharpness company: The Trust appointed as their executive engineer, under Keeling, E. J. Cullis, a man of 36 and an associate member of the Institution of Civil Engineers, who had been resident engineer at the Gloucester Docks.[40] W. J. Snape of the Stroudwater acted as secretary. As the existence of the old company was preserved in the interests of the contingent annuitants, a new seal had to be designed for the Trust: in contrast to the original it was severely plain.

Of the £15,000 borrowed, £4,688 were spent in discharging the debentures and meeting parliamentary expenses, so that there remained for reconstruction of the canal only a few odd pounds more than the estimate — nothing, that is, for treating those unknown points of weakness the engineers were aware might so easily exist. Plenty of these were soon discovered, and so, at the beginning of 1898, although the canal had been re-opened to Daneway and the tunnel made secure, there remained in the eastern part of the summit leaks to be stopped, dredging to be done, lock gates to be repaired — and a balance which was wholly inadequate. It was reckoned that another £1,297 would be needed. About this time Cullis resigned. Tributes paid at the time of his death more than thirty years later show him to have been a man of integrity and strong character, highly esteemed professionally. So, in the absence of documentary evidence of the Trust's view, it seems probable that Cullis quitted their service because he was dissatisfied both with what he had seen and what he foresaw. Be that as it may, from August 1898 Snape was in sole charge locally, but although he had long experience of canal management and maintenance, he was not, as Cullis was, a trained engineer. To meet their immediate financial needs, the Trust decided to ask for the loan of another £2,000. The Wilts & Berks was by then incapable of shouldering any additional burden. The Severn Commissioners concluded it would be 'ultra vires' for them to do so. But the other navigations and public bodies, already so deeply committed, agreed to guarantee the interest on this additional loan.[41]

The work was done, but only by doubling the loan, and on 15 March 1899 the canal was opened throughout once again. Charles Hooper & Sons, mill-owners at Eastington (Charles Hooper represented the Stroudwater on the Trust's committee), chartered a narrow boat from a well-known local owner, James Smart of Chalford, to fetch a cargo of wool from London. After about three weeks, *The Trial* returned with 92 bales of colonial wool, and on 11 April was photographed as she passed triumphantly between the Walbridge locks. The local press eulogised this "practical demonstration of the re-opening of the communication by water between east and west".[42]

Alas! it needed but a few short weeks of summer to shatter the illusion of success. The summit was not watertight. Navigation ceased in June. Through summer and autumn "numerous leaks" were stopped, but at the end of October the Trust's committee confessed that between Trewsbury Bridge and the pumping station the summit still leaked badly. This was the principal point of weakness noted in 1894 when the cost of re-lining a little more than half-a-mile had been estimated at £1,660, yet Keeling now advised that it would have to be "entirely re-puddled — and in some places the bottom concreted — for a distance of about a mile and a quarter" at a cost of between two and three thousand pounds. After all that had been done — and the expenditure, not of £10,309, but of more than £14,200 — the committee could claim no greater achievement than to have put the rest of the canal "in fair order". Further work they could not undertake, for the Trust having already exceeded its statutory borrowing powers, the committee were in a ticklish position and certainly could not raise another loan without the authority of Parliament.[43]

Why had the Trust failed so ignominiously? One published account remarked that the work was so badly done that the money was wasted.[44]

Others commented that lengths re-puddled were not immediately filled with water, so that the puddle dried and cracked. Sir William Marling himself said that the Trust never had enough money (that, though he did not say it, was due to the faulty estimates) and so had been driven to patch when they ought to have reconstructed.[45] Essentially, however, the failure derived from the unwillingness of the engineers to respect the opinions of the past. Although they could not easily have found the letters in which Denyer and Charles Jones reported how the springs in the great oolite blew up the puddle in winter, they had plenty of other evidence. In 1877 Whitworth's warning had been quoted and scornfully rejected; in 1883 Taunton had mentioned McIntosh's report of 1820 that the nature of the ground rendered periodic leakage inevitable; in 1894 they showed their knowledge of Taunton's statement that through the great oolite near Coates there was no way of keeping the canal permanently watertight.[46]

There is no escape from the conclusion that the Allied Navigations' scheme was an ill-advised one, devised in ignorance, developed in over-confidence, and promoted out of pique. Their officers could learn only through bitter experience. By the end of 1898 they had done so. In December an article was published in *The Engineer* remarking of the Thames & Severn, summit: "it is doubtful whether there is any portion of canal in England more troublesome to keep water-tight than this has been."[47]

As the Trust's position was anomalous, it was more necessary than ever before to pass the buck.

REFERENCES

1. T & S Collection. GWR Minute Book of the Directors, 22 February 1883. GWR half-yearly reports 1882-94 show the position of Holland and Mortimer.
2. SND & G & B Minutes 7 June 1882 reveals the Sharpness board's suspicion. GWR Minute Book of the Directors, 21 February 1883 records that Stanton had written about the condition of the T & S, but as the GWR directors forwarded the letter to "the Managers of the Canal Co.", this particular 'feeler' provided no confirmation.
3. Report from the Select Committee on Canals 1883, pp. 103-4.
4. SND & G & B Minutes, 1 November 1882. GWR Minute Book of the Directors, 16 November 1882. Letter from J. H. Taunton to Henry Waddy 2 December 1882 (T & S Collection).
5. *Report to the Allied Navigations re. Thames & Severn Canal* 1894. The report made by Clegram, Marten and Snape in 1883 is reprinted on pp. 22-4.

6. PP 1886 Accounts & Papers vol. 160, pp. 700, 705-7. PP 1887 Accounts & Papers vol. 72, p. 889.
7. PP 1886 Accounts & Papers vol. 60, pp. 703, 705.
8. PP 1887 Accounts & Papers vol. 72, p. 888.
9. *Thames & Severn Canal Navigation, Report of Proceedings at a Meeting of the Shareholders held at Reading 16 January 1894*, p. 5 (Printed).
10. PP 1887 Accounts & Papers vol. 72, pp. 886, 893.
11. PP 1888 Accounts & Papers vol. 89, p. 577; Taunton's report of 4 July 1885.
12. PP 1886 Accounts & Papers vol. 60, p. 702, letter of 21 May 1886.
13. PP 1887 Accounts & Papers vol. 72, pp. 886-91, 894. PP 1886 Accounts & Papers vol. 60, pp. 695, 702.
14. PP 1886 Accounts & Papers vo1. 60, pp. 691-710, and PP 1887 Accounts & Papers vol. 72, pp. 883-97 covering the period 4 March 1886 to 29 January 1887. Cf SND & G & B Minutes 5 August 1885, 6 January, 7 April 1886.
15. PP 1887 Accounts & Papers vol. 72, p. 895.
16. Ibid pp. 888, 890.
17. Ibid pp. 890, 897.
18. PP 1888 Accounts & Papers vol. 89, pp. 575-7.
19. SWN committee, 21 August, 23 October, 18 December 1888, 19 March 1889.
20. Cf Proceedings Commons Committee on Thames & Severn Railways Bill 1866, evidence of J. H. Taunton that of 18 barge owners petitioning against the Bill, only 10 had carried anything on the canal in 1864 and only 9 in 1885 (T & S Collection).
21. Thames & Severn Canal, In Parliament, Session 1895, Statement as to the means of acquiring and maintaining this canal (GCL HF.5.11). Gardom op cit pp. 7-14.
22. Undated extract from *Wilts & Gloucestershire Standard*, early January 1894.
23. PP 1888 Accounts & Papers vol. 89, p. 576.
24. Undated extract from *Wilts & Gloucestershire Standard* early January 1894.
25. See for example an extract from the same February 1895.
26. Thames & Severn Canal, In Parliament, Session 1895 . . . Gardom op cit pp. 9-13. The reports of the speeches quoted in both were taken from the *Stroud Journal*, 12 January 1894.
27. SND & G & B Minutes, 3 August 1887.
28. Gardom op cit, p. 12.
29. Pratt, *British Canals*, 1906, gives many instances of railway expenditure to cover canal revenue and maintenance in the preceding year.

30. Gardom op cit, pp. 12-13.
31. *Wilts & Gloucestershire Standard* 13 January 1894.
32. SND & G & B Minutes, 6 September 1882.
33. GJ, 31 October 1903, report of a speech at a meeting of the Gloucestershire County Council by R. A. Lister who had made private enquiries at Paddington concerning the steps taken by the GWR authorities before deciding to give up the T & S Canal.
34. *Thames & Severn Canal Navigation, Report of Proceedings at a Meeting of the Shareholders held at Reading 16 January 1894*, pp. 8, 11, 16, 18.
35. *Report to the Allied Navigations re. Thames & Severn Canal* 1894.
36. Notes of a meeting held at the Great Western Royal Hotel, Paddington, 2 November 1894, and Proposals of 29 January 1895 (T & S Collection).
37. Act 58 & 59 Vict. cap 149.
38. Gardom op cit p. 20.
39. Memorandum of an inspection of the River Thames between Lechlade and Oxford, 2 April 1895 (GCL). Gardom op cit pp. 4, 21. Thacker op cit vol. 2, pp. 50, 59, 74, 84.
40. I am indebted to the Secretary of the Institution of Civil Engineers for details of the early part of the career of E. J. Cullis. Other information about him is drawn from the obituary notice and tribute in GJ, 12 April 1930, and SND & G & B Minutes 14, 28 August, 9 October 1895.
41. Letter from Sir William Marling to the sponsoring bodies c. 14 January 1898. SWN committee, 20 April, 18 May, 21 December 1897. Gardom op cit p. 19. *The Engineer*, 8 May 1896, 23 December 1898.
42. Printed leaflet "*April, 1899. From London to Eastington by Water*". A copy is pasted in The Fact Book (T & S Collection).
43. Gardom op cit pp. 21-2. The County Council's T & S balance sheet 30 September 1914 (GCL), records the exact expenditure of the Trust as £18,890 16s. 1d., from which must be deducted £4,688 10s. absorbed by the debentures and parliamentary expenses.
44. *The Municipal Journal*, 30 August 1907, pp. 741-2 "Gloucestershire's Municipal Canal".
45. GJ, 13 January 1900, report of meeting of the County Council.
46. Report from the Select Committee on Canals 1883, p. 114. *Report to the Allied Navigations re. Thames & Severn Canal*, p. 21.
47. *The Engineer*, 23 December 1898.

CHAPTER FOURTEEN

THE DEATH AGONIES
II: COUNCILLORS ADRIFT

Sir William Marling put the position to the chairman of the Gloucestershire County Council,[1] guarantor of the largest single contribution to the funds. The chairman was Sir John Dorington, who had been a member of the old company's committee of management for several years nearly a quarter of a century before. He was ready to propose that the County Council should take on where the Trust perforce left off. But, it was said later,[2] that Sir William was so anxious to extricate the Trust from its embarrassing position by shifting the liability to the shoulders of the ratepayers that in his dealings with the County Council he failed to speak frankly of the engineering problems, of whose magnitude he was by that time well aware. It is a curious fact that not even the minute books of that period passed into the possession of the County and that they have never been traced.

When Sir John Dorington put his proposition to the County Council, he said that it was "perhaps the most important matter which had ever been brought before" them.[3] At that date, it was indeed what the *Gloucester Journal* later called a "daring adventure into the realm of Municipal Socialism".[4] He had already obtained an independent estimate for treating the leaking reach of the summit, and in both extent and cost (1½ miles, £5,000) it exceeded Keeling's. Nevertheless it appeared reliable because it had been prepared by G. R. Jebb, engineer to the Shropshire Union Railways & Canal Company which was considered one of the best managed and most enterprising canal systems in the country[5] (it was leased by the LNWR who made extensive use of it to reach areas not served by their own metals). Sir John Dorington expected restoration to be completed in three or four months, and urged the Council to undertake the work rather than sacrifice the money already spent, which the County had in any case to redeem over a period of years.

No one really opposed the proposition — so few had any valid experience on which to form an opinion. One who had was Councillor Peter from Berkeley, directly involved in the maintenance of the sea walls along the Severn estuary, but it was not until the next meeting that he was able to warn the council that, having spent several thousands on the canal, "they would very likely find" that they had to spend as much again.[6] Meantime a note of caution was sounded by R. A. Lister, the able and far-sighted founder of the Dursley engineering firm, who doubted the efficiency of public ownership in general and mistrusted the accuracy even of Jebb's figures because "the cost of repairs in connection with waterways, . . . mill-ponds, dams, etc. . . . was almost impossible to estimate with any degree of correctness". He thought the Council might well find themselves committed to an expenditure of £10,000 or more. Local politics no doubt coloured the debate to some extent (Dorington was a Conservative, Lister a Liberal), but even Lister conceded that "it was undoubtedly important" to keep the canal open as a route in competition with the railway, a point the *Gloucester Journal* had already mentioned .[7] There was only one member with real knowledge of what was involved, Sir William Marling, and Lister, like others, therefore looked to him for an informed opinion. But when Marling spoke, not all his remarks were audible, and those the reporters of the *Journal* were able to record showed him more concerned to excuse the failure of the Trust than to offer constructive advice. It was therefore left to Sir John Dorington to dismiss Lister's forebodings. He did so scornfully, and the Council, influenced no less by Sir John's advocacy of the canal than by his popularity, and great experience of local government, unanimously approved his proposition.

And so on 8 January 1900 the young Gloucestershire County Council embarked upon a brave but utterly disastrous enterprise. They followed the procedure suggested by the MP for Stroud during the meeting with the President of the Board of Trade in 1894: applied for a warrant authorising the Trust to abandon the canal on the ground that it had become derelict, and for a provisional order transferring it to the County Council. This was, it appears, the only case in which that procedure was adopted.[8]

The warrant was issued on 30 June, but confirmation of the order, assumption of the Trust's debts, and authority to raise a loan, required the consent of Parliament. It was very important that the works should be resumed without delay, but a Bill could only be introduced that session if the standing orders of Parliament should be suspended. Notice was given and there appeared to be no serious objection, but the Council had failed to reckon with Michael Biddulph MP who had inherited the Kemble estate from Robert Gordon's daughter.[9] When the Bill was introduced in the Lords, objection was raised on his behalf to the abstraction of "his water" — that water for which, as the Council knew full well, Robert Gordon had been paid £1,000 nearly seventy years before. This was a set back. The Council authorised execution of "some trifling work",[10] but 1900 passed and the summer of 1901 came, before Parliament confirmed the order and they could do more. The Act authorised the Council to borrow £34,000, £19,000 of it to cover the Trust's debts, and confirmed the liability of the Allied Navigations and other public bodies to maintain their previous contributions.[11]

To manage the canal, the County Council appointed a committee of ten, including Sir John Dorington, G. B. Witts (a civil engineer), R. A. Lister, Sir William Marling, and Philip Evans. Evans was especially active, and in course of time earned tributes from his colleagues for "the enormous amount of work" that he personally did towards the restoration.[12] All commercial business was handled by W. J. Snape in addition to his work for the Stroudwater Navigation, and for this the Council paid him £100 a year with 30s. a week for an assistant. The county surveyor had been expected to superintend the engineering, but as he pleaded illness and pressure of other work, the committee invited, on Witts's recommendation, A. Brome Wilson, a young civil engineer of Cheltenham, to do so. Wilson asked £25 a month, rather more than the committee had expected to pay, though no more than Clowes had received over a century earlier.

Wilson got to the point at once. Even before his appointment had been confirmed, he studied the summit and talked to "gentlemen who [had] known the Canal for many years" as well as "labouring men who [had] worked upon it".[13] It is quite clear it was the information he derived from these talks which enabled him to re-establish so quickly that chain of cause and effect which Denyer and Jones had observed many years before, and that it was this knowledge that made his appreciation of the situation so much more realistic than that of any of the other engineers recently involved. In Wilson's opinion, the critical length was far more extensive than either Keeling or Jebb had supposed, and extended more than three miles. Most of this could be made secure by stripping and re-puddling, but not so the deep cutting east of the tunnel, and it was Wilson's comment upon this reach that led Dorington to tell the next meeting of the council that "a very serious engineering difficulty had presented itself", as "the whole of the bottom of the canal for a quarter of a mile" was "honeycombed by springs".[14] The cost of treating this section alone was expected to be £3,000 and before allowing anything to be done, Dorington proposed to consult Sir Benjamin Baker, whose fame as engineer of the Forth Bridge had set him at the peak of his profession.

Baker approved the young engineer's plan[15] for tracing all the fissures in the rock, covering them with a concrete invert 18 inches thick, and inserting pipes which would lead water from the springs to discharge into the canal well above the water-line. Using direct labour, Wilson concreted the deep cutting from end to end with entire success, but the committee decided that re-puddling of the remainder of the three mile length should be done by contract. The tenders gave the councillors another shock, but they decided to accept the offer of Braithwaite & Co. of Leeds to do the work for £12,000.

Dorington's estimate of the total cost of restoration had now risen to £17,000, but Lister, whose membership of the committee never stopped his well-founded criticism of the whole enterprise, reckoned it would be at least £20,000 and added that he would "not be at all surprised" if it eventually reached £30,000. There were cries of "No, no!", a comment from Marling that £30,000 "was a little wild", and the council remained unconvinced.[16]

Braithwaites puddled for nine months with little success. Wilson found them using clay in lumps too large, adding too much water, failing to work the material into the essential homogeneous mass, and reported to his committee (with a faint echo of James Perry's words) "that the work was being scamped".[17] Braithwaites were made to rework a part of it, and given notice to quit. Claim and counter-claim were subsequently lodged before the courts but then withdrawn. Wilson himself took over the puddling.

Much other work was also done: Smerrill Aqueduct re-puddled; basins and side-ponds re-lined with clay or concrete; bulges removed and brickwork repaired in the tunnel, and parts of the bottom lined with concrete to exclude the springs. Seeking more water, the County Council even plugged holes made by vermin in Allen's Millpond (once Jenour's) and the swilly-holes in the bed of the River Churn. There was a great deal of dredging to be done, and after four men had shifted 1,200 tons of mud with a hand-dredger, the committee bought a second-hand steam dredger for £138. Named *Empress*, she had to

Empress dredging at Chalford in the early 1900s.

Staunch discharging coal at Cirencester, 19 March 1904.

voyage from Bath to Brimscombe via Oxford, Birmingham and Gloucester, pausing at Oxford while Salter Bros. repaired her hull and at Gloucester for Sisson & Co. to fit a new boiler; then she was re-decked for'ard and, to make her suitable for use along the lonely Cotswold reaches, given a cover over engine and boiler and a cabin fitted with bunks. Equipped with several silt-boats, *Empress* then dredged the Thames & Severn Canal for about a dozen years.[18]

Stage by stage, the canal was reopened: Cirencester to the Thames by July 1902; from the west to Daneway by April 1903; the remainder of the summit by January 1904. But the only junketings were the resumption of those hilarious excursions beloved of some Cirencester folk, by boat to Tunnel House — sometimes weaving their way back in darkness with an unsteady hand upon the tiller — and the decorous applause in the council chamber. Meantime Dorington's 'three or four months' became as many years, and when the restoration account was closed, the total (excluding the Trust's debts) was £29,926,19 proving Lister's 'rather wild' estimate a remarkably accurate one. Even so, imperfections remained, as Brome Wilson made clear in the final report he submitted before relinquishing his task in April 1905. Although he made some mistakes, for which Lister criticised him, the evidence strongly suggests that Wilson emerged from the final ordeal of the Thames &

Severn Canal with more credit than any of the other engineers concerned, and the committee's vote of thanks and his own appreciation of the "kindness, consideration, and courtesy" shown him by the full Council suggest that in general relations had been harmonious and confident mutual.[20]

The first vessel passed over the reopened summit on 3 March 1904. She was the horse-drawn narrow boat *Staunch*, belonging to Joseph Hewer, whose history shows the precarious livelihood of a western boatman around the turn of the century. Born at Chalford and reared by his father on narrow boats from the age of seven, he migrated to Swindon about 1890 because of the condition of the Thames & Severn, and for a dozen years or more had four boats plying regularly between Swindon and Bristol. When the Wilts & Berks also became un-navigable, he was attracted homeward by news of the restoration works in progress on the Thames & Severn. Breaking up two boats and selling a third, he came only with *Staunch* and waited at Latton for ten months before he could get any further. After carrying a few cargoes on the eastern reaches while waiting for the summit to reopen, he set out for a Staffordshire colliery, picking up a cargo of timber from Gloucester to Old Hill, Dudley, en route. At Hednesford Colliery he took on board some 37 tons of coal for Gegg & Co. of Cirencester. Leaving on a Wednesday morning with more than 120 miles to cover, and lying up near Stroud over the Sunday, he tied up at Cirencester Wharf at 10 am on Tuesday 19 March with the first cargo of waterborne coal to reach that port for sixteen years. As he had charged 7s. 6d. per ton, the same as the railway rate, he earned £13 17s. 6d., out of which he paid between £5 and £6 in navigation tolls.[21] Allowing for labour, food, horse-keep, and depreciation of the boat, the margin of profit cannot have been large.

Various pleasure craft traversed the canal on their way to or from the Thames, a type of traffic Sir John Dorington rightly claimed had "never existed before",[22] for previous managements had never encouraged it. P. Bonthron described one such journey in *My Holidays on Inland Waterways*. Setting out from Chertsey in the motor launch *Balgonie*, he travelled up the Thames and via Stroud, Gloucester, Birmingham and Oxford, but having reached Cirencester at Whitsun he was delayed for a week by the traditional closure for repair. Warned there of weed in the canal ahead, he took the precaution of hiring a horse and a man who indeed were needed to tow the launch all the way to the tunnel.[23] E. Temple Thurston came in the *Flower of Gloster*, a narrow boat whose lovely name is the title of the most charming of all such books. *The Flower of Gloster* had once plied on the Thames & Severn, but, like Hewer, she had been forced to move, and Temple Thurston had hired her at Oxford together with a horse and an endearing character named Eynsham Harry with whom he reached the edge of the Black Country before returning via Framilode and Inglesham.[24]

Through Sapperton Tunnel, Bonthron and his friends took *Balgonie* in

A party of excursionists emerging from the tunnel at Daneway in the early 1900s.

A trow in the south-west corner of Brimscombe Port in the early 1900s.

ninety-five minutes, pulling with boat hooks on a wire fixed to the wall. Temple Thurston and Eynsham Harry did it in the time-honoured way, lying on the wings and legging against the side walls for four hours while water dripped on their faces. Hewer took *Staunch* through in 3¼ hours, he and his mate propelling her "by means of tunnel sticks" against the walls, a method the old company had strictly forbidden for fear of damage to the brickwork.

Trade returned to the canal but slowly. After all, except for *The Trial*, scarcely a boat had passed the summit for ten years or more before *Staunch*, and other boatmen than Hewer had had to migrate in search of a living.[25] Nor was there much incentive for them to return as long as the County Council held interminable quarterly debates in which speakers questioned the wisdom of continuing to maintain the summit in navigable condition. Some members of the canal committee indulged in wishful thinking, and, eager to justify the works, sought to impress the councillors with facts and figures which would not bear examination. Dorington's expectation of "an enormous pleasure traffic" amounted to perhaps half-a-dozen boats a year.[26] The thousand quarters of barley which "he was rather pleased to hear" would be shipped from Berkshire to Stroud filled perhaps half a score of boats.[27] The contract with "a certain brickworks, . . . which had not at one time contemplated sending anything by the canal",[28] was revealed by Lister as no more than the carriage from Stonehouse to Kempsford of 17,000 bricks — "about enough to build a good-sized cottage" — and worth 48s. 7d. tolls at the most.[29]

It was all rather pathetic. Lister, inspecting the canal three times in part and twice from end to end, saw "scarcely any evidence of traffic" as he walked for miles along the towing-path.[30] Temple Thurston met only one boat, and that in the Golden Valley. Though Evans claimed that the proportion of boats passing beyond Chalford showed "the bulk of the traffic was of a through nature",[31] only eighty used the summit in the first twelve months, and the plain facts are that in the four years to September 1908, 82 per cent of the traffic and 63 per cent of the earnings were confined to Chalford or below. During that same period the average income from tolls was no more than £298, about a quarter of the cost of maintenance, and the annual loss was £3,618, all but £975 of which had to be found by the Gloucestershire County Council.[32]

No wonder Lister's persistent probing increased in force and frequency as the years passed. No wonder he twice asserted that a new chairman of the Council would drop the canal.[33] In 1907, he made its maintenance an issue in the County Council election,[34] and when the new Council met, introduced a motion proposing to close the canal beyond Chalford. The time was still not ripe, but doubts of the chairman's policy had spread so far that, when Lister's motion was put to the vote, it was only defeated by 35 to 23. It could not even be claimed that the restoration had been successful, for in 1905, 1906,

Navvies at Puck Mill in 1907.

Puddling Puck Mill pound in 1907.

1907 and 1908, the canal was closed to through traffic for twelve weeks or more while further repairs were executed — on the summit except in 1907 when Puck Mill pound, leaking so badly that it "emptied itself about every four hours",[35] was re-puddled. In the spring of 1908, Dorington resigned the chair, and in the autumn, at last acknowledging the Council's inability to keep the canal open throughout the year, he moved a resolution appointing a special committee to examine the position.

It is a sobering thought that a succession of professional engineers had failed where the practical experience of Denyer and Jones had succeeded many years earlier.

Among the fifteen members of the special committee were Dorington, Lister, Evans, Marling, and Lord St Aldwyn, the chairman, a former President of the Board of Trade and Chancellor of the Exchequer. There was only one conclusion they could reach, but after all the public money which had been sunk, they were understandably reluctant to reach it.[36] Asked for their advice, Snape and Keeling concurred in the opinion that £2,500 would soon have to be spent in renewals and repairs. The special committee decided to approach the Thames Conservancy, hoping to persuade them to take the canal off the Council's hands. First reactions were favourable, but the Conservancy wisely hesitated before committing themselves, and the negotiations were protracted. Eventually it was agreed that the Conservancy's engineer should inspect the canal, and this — with the exception of the tunnel — he did in November 1909. In his opinion, the expenditure of £8,234 would be necessary almost immediately, but although this was more than three times Snape's and Keeling's figure, it was only a beginning. He reported bluntly that as clay puddle had so repeatedly proved ineffective against the fissures and springs, many parts would have to be lined with concrete at a cost of £24,500. Furthermore, as the seasonal variation in the water level was so great, it might well prove necessary to build concrete retaining walls throughout at the cost of a further £72,864.[37] It was high time someone took a realistic view, but his total of more than £100,000, excluding repair of the tunnel, was staggering. He advised the Conservancy not to take over the canal, and who dare say he was wrong?

So the County Council had no chance to pass the buck.

Lord St Aldwyn had changed his opinion of the canal's value-to the community several times: in 1866, supporting the Thames & Severn Railways Bill, he considered the waterway of little use;[38] in 1894, he told Mundella that its closure "would be a very great injury to the district";[39] but he now called it a "white elephant" which the County Council "would find it extremely difficult to get rid of", and quite gratuitously admonished them for having exceeded the duties committed to them by Parliament in taking it over.[40] The 'white elephant' was indeed difficult to get rid of. There were problems of long-standing water rights, disposal of flood waters, maintenance of bridges, of ensuring that the navigations and public bodies remained liable to pay their annual contributions until the loans had been redeemed.[41] The special committee therefore proceeded cautiously and, instead of proposing complete abandonment, recommended that no more should be spent on the canal east of Chalford.

Even so, the obsequies took a long time. Water was pumped into the summit in September 1910. A boat with 24 tons of gram passed over it to Kempsford and Lechlade in April 1911, and another the last with 20 tons of stone on 11 May 1911. Almost all the men employed east of Chalford were discharged in 1912, but a few repairs were done thereafter, and even as late as 1916 arrangements were made to insert struts in the tunnel after two falls of the roof.

In June 1920 it was "reported that there were practically no boats left in the district",[42] and towards the end of 1923 tentative steps were taken so that the whole canal might be abandoned in fifteen months' time when all the loans would have been repaid. Inevitably there was local opposition, principally from the Stroudwater company, who benefited not only from trade passing to and from wharves higher up the valley, but also from a valuable supply of water through the Walbridge locks. The Ministry of Transport therefore held an enquiry, after which the Stroudwater agreed to pay £770 a year towards upkeep of the lower reaches of the Thames & Severn and to withdraw opposition to abandonment of the remainder. Thus it was 31 January 1927

Fall of the tunnel roof in the Long Arching through the fuller's earth, photographed by C. P. Weaver in 1955.

Hope Mill in 1948.

before abandonment between Lechlade and Whitehall Bridge was authorised. By that date, the total cost, including restoration carried out by the Trust and the County, payment of interest, and losses incurred in working, amounted to about £64,000, not all of which of course had fallen on the ratepayers of Gloucestershire, though theirs had been the major burden.

At the end of that decade, some eighty or ninety boats were using the canal annually, but there was a sharp drop to about 25 in the early thirties as road motor competition increased, and by then what remained of the canal was really only of value to the boat building establishments at The Bourn and Hope Mill. The Bourn had been leased by several generations of the Gardiner family who had built Stroud barges and other craft in the dry docks originally made by the Thames & Severn company. The industry at Hope Mill was more recent and more enterprising. In 1878 Edwin Clarke & Co. had established slipways and engineering workshops where they and their successors, Abdela & Mitchell Ltd., built launches and tugs, cargo and passenger river vessels, fitted with screws or stern paddle-wheels driven by steam or diesel. Some of these boats had gone through the canal to Inglesham and down the Thames to London for shipment abroad. Others, too large to pass the locks, were shipped in parts for assembly abroad, or built in sections which were floated independently down the canals to the junction with the Gloucester & Berkeley at Saul, where they were joined together. The firm struggled on through the depression of the early thirties, and even for a few years following the abandonment of the lower portion of the canal.[43]

This was authorised by order of 9 June 1933, after which the lands and buildings were sold, and the canal ceased to be maintained except as a channel for the escape of floodwaters. After a century-and-a-half during which no locked gates, warning notices, or actions for trespass had hindered public access, the towing-path had undoubtedly become a public right-of-way.[44] The right has only been exercised over short isolated lengths east of the tunnel, where the path in many places has become densely overgrown, but it is maintained throughout the Golden Valley, and it is to be hoped that the right to a walk of such beauty as that between Chalford and Daneway will long be cherished.

The Stroudwater Navigation struggled on for twenty years. The last dividend was paid in 1922, by 1924 the canal was working at a loss, and in 1941 all traffic ceased. Even so, it was long before the navigation was officially abandoned. The first tentative steps were opposed by the Gloucestershire County Council, who claimed that the Stroudwater was the natural outlet for water still flowing from the Thames & Severn. But after the Stroudwater had lodged a suit in Chancery, the two parties reached agreement: the Council accepted they had no right to such use of the navigation, that they would maintain bridges carrying County roads, and would withdraw their opposition on condition they should be protected from any liability for the abandoned waterway.[45]

The last use of the canal? Bathing in Griffin's lock during the long hot summer of 1947.

The way then seemed clear, and notice of abandonment was given on 17 January 1952. After more than ten years of disuse, objection seemed unlikely, But romantic visions were evoked, a preservation society was formed, and the Inland Waterways Association, which had done so much to revive interest in canals, whipped up opposition. Unhappily, their attitude was no more realistic than that of the Allied Navigations more than fifty years earlier: closure of the Stroudwater would destroy hope of reviving the Thames & Severn, "which, in spite of its troubled history, could survive and pay its way", as "although it had its own difficulties, its failure was due to the state of the Upper Thames"[46] Nevertheless, saner counsels prevailed and the Act authorising abandonment was passed in July 1954.[47]

The Stroudwater company has survived as owners of property and vendors of water. A careful corporation, they had always preserved the privacy of their towing-path, so no public footpath runs through the fields between Stroud and the Severn. But pleasant though it is, a canal-side path is nearly always the longest of all possible routes between two points.

Since then an active and enthusiastic Stroudwater Thames & Severn Canal Trust has been formed with the avowed intention of restoring navigation. The problems confronting them have, of course, been greatly increased by alienation of the property: in the 'populous Vale', much, including the site of Brimscombe Port, has been built over, elsewhere parts of the bed have been incorporated in local landholdings and hump-backed bridges have been replaced by level arches. But the Trust's view has been realistic, and in 1976 they commissioned a well-known firm of consulting engineers to examine the line and report on the feasibility of their project. This was done with great thoroughness, including close inspection of the tunnel, and the conclusion was that navigation could be restored at a cost of about eight million pounds.[48] That is a lot of money to find.

Meantime, the Trust has done admirable work in saving that notable monument of the canal age, the eastern end of Sapperton Tunnel, and also in clearing and restoring a deep cutting approaching it. So, standing on Tarlton Bridge, there is a vista over water as far as the portal set against the backdrop of Hailey Wood, and as Tunnel House has been rebuilt (although on a reduced scale), this is a very pleasant spot to visit.

REFERENCES

The material in this chapter has been drawn from the Minutes of the Canal Committee of the Gloucestershire County Council and other sources in the County Records Office (Thames & Severn Collection), and from the informative files of the *Gloucester Journal*, 1900-1910 wherein the quarterly debates of the County Council, in which the affairs of the canal figured prominently, are reported at great length. Detailed references are given only where material has been drawn from other sources, or passages have been quoted from the *Journal*.

1. Gardom op cit, p. 22.
2. See GJ, 2 March 1907, letters from Sir William Marling and R. A. Lister, and 14 November 1903, Lister's remarks at a meeting of the County Council.
3. GJ, 13 January 1900.
4. GJ, 23 April 1910.
5. Royal Commission on Canals & Waterways 1906-1909, vol. 1 part 2, item 4131 evidence of E. C. Corbett, "I think it is the best managed canal in England".
6. GJ, 14 April 1900.
7. GJ, 13 January 1900.
8. Royal Commission on Canals & Waterways 1906-1909, vol. 7, p. 14, section 61.
9. Kemble Women's Institute, *A Guide to Kemble & Ewen*, 1951, p. 7.
10. GJ, 12 January 1901.
11. Act 1, Edw. VII cap 3.
12. GJ, 31 October 1903.
13. Gardom op cit, Wilson's report of 12 April 1901 is quoted in full on pp. 28-32.
14. GJ, 20 April 1901.
15. Gardom op cit, Sir Benjamin Baker's report is quoted in full on pp. 33-35.
16. GJ, 5 October, 1901.
17. GCC Canal Committee Minutes, 7 April 1903.
18. Report by W. J. Snape, 13 December 1902 (GCL JF.14.38). See also Minutes and Journal files.
19. GCC T & S balance sheet, 30 September 1914 (GCL HF.5.37): the restoration account had been closed in 1905.
20. GJ, 16 January 1904.
21. *Wilts & Gloucestershire Standard*, 26 March 1904.
22. GJ, 5 October 1901.
23. Bonthron, P., *My Holidays on Inland Waterways*, 1916, pp. 2-3, 7-10.
24. Temple Thurston, E., *The Flower of Gloster*, 1913, pp. 208-244.
25. For example, Gleed of Chalford who told me that he had moved to Staffordshire for a time.
26. GJ, 5 October 1901. The Minutes record 13 in 1908 and 1909.
27. GJ, 12 January 1907.

28. GJ, 16 April 1904.
29. GJ, 13 July 1907.
30. GJ, 13 July 1907.
31. GJ, 8 July 1905. There was, and is, a good deal of confusion surrounding the number of boats which actually used the summit; Evans certainly once included those trading to Daneway which normally would not even have passed the summit lock. Lister, who is not likely to have exaggerated, gave the figure of 80 in the first year (GJ, 14 January 1905). Griffiths, the Thames Conservancy's engineer, wrote that it never exceeded 46 in any one year (GCL JF.14.47). The minutes record most, though not all, of the quarterly returns, and these certainly show 100 boats passing over the summit in the year ending 30 September 1908.
32. In the minutes is a summary of tons carried and tolls earned, and a description of freights carried, for the four years ended 30 September 1908. A statement by Sir John Dorrington 13 February 1909, covering the same period, shows a 'Trade Deficit' which was really the difference between the income from tolls and other sources and the expenditure on maintenance (T & S Collection).
33. See GJ, 31 October 1903 and 12 January 1907.
34. See GJ, 13 July 1907 Local Notes.
35. GJ, 13 July 1907.
36. See statement by Viscount St Aldwyn, 13 March 1909 (GCL J.14.11).
37. Thames Conservancy. Report on the Thames & Severn Canal, March 1910 (GCL JF.14.47).
38. Proceedings Commons Committee on Thames & Severn Railways Bill 1866, evidence of Sir M. E. H. Beach (created Viscount St Aldwyn in 1906).
39. Gardom op cit, p. 9.
40. GJ, 23 April 1910.
41. P. J. Evans expressed the opinion that the contributions would probably not be forthcoming if the lower part of the canal should be closed — see statement on the finances of the canal, 11 July 1912 (GCL JF.14.45).
42. GCC Canal Committee Minutes, 25 June 1920.
43. Correspondence, specifications, and photographs found at Hope Mill and now in the County Records Office.
44. See report of the Clerk of the County Council on footpath rights, 1 July 1935 (T & S Collection).
45. *Stroud Journal*, 4 May 1951. Minutes of a sub-committee of the GCC. SWN records.
46. Inland Waterways Association *Bulletin*, 33 August 1952.
47. *Stroud Journal*, 15 January 1954. Gloucester *Citizen*, 15 January 1954.
48. Freeman Fox Braine & Partners: *Preliminary Engineering Report on the Feasibility of Restoring the Canal to Through Navigation*, August 1976.

APPENDIX 1

TABLE OF LOCKS

The Stroudwater Navigation rises from low water in the River Severn, which is 14 feet 5 inches above ordnance datum, to Walbridge Basin, Stroud, 107 feet 2 inches.

The Thames & Severn Canal rises 241 feet from Walbridge to the summit, 348 feet 1 inch above low water River Severn and 362 feet 6 inches above ordnance datum.

	Distance from Walbridge		Rise as recorded in 1810	
	Miles	Chains	Feet	Inches
Locks 68 to 69 feet long, 16 feet 1 inch or 16 feet 2 inches wide				
Walbridge Lower	0	0	90	
Walbridge Upper	0	11	11	0
Bowbridge	0	71	10	0
Griffin's	1	22	10	0
Ham Mill	1	50	9	0 a
Ridler's or Hope Mill	2	1	8	0
Gough's Orchard or Dallaway's or Lewis's	2	20	8	0
Lock 90 feet long and 16 feet 1 inch wide				
Bourn or Harris's	2	61	11	0
Locks 90 to 93 feet long and 12 feet 9 inches or 13 feet wide				
Beale's	3	21	8	0
St Mary's or Clark's	3	52	8	0
Grist Mill or Iles's Mill or Wallbank's	3	60	8	0
Ballinger's	3	70	8	0
Chalford Chapel	4	2	8	0
Bell	4	25	10	0
Innell's or Clowes's or Red Lion	4	40	8	0
Golden Valley	4	62	8	0
Baker's Mill or Twizzel's Mill Lower, later Bolting	5	36	8	0

Baker's Mill or Twizzel's Mill Upper, or just Baker's Mill	5	44	8	0
Puck Mill Lower	5	64	8	0
Puck Mill Upper	5	70	8	0
Whitehall Lower	6	10	8	0
Whitehall Upper	6	50	8	0*
Bathurst's Meadow or Bathurst's Meadow Lower	6	60	8	0*
Bathurst's Meadow Upper or Sickeridge Wood Lower	6	67	8	0*
Sickeridge Wood Middle	6	72	8	0*
Sickeridge Wood Upper	7	0	9	0*
Daneway Basin or Daneway Lower	7	5	9	0*
Daneway Bridge or Daneway Upper	7	10	9	0*
Fall from the summit to the Thames 129 feet				
Siddington Upper	15	23	9	9
Siddington Second	15	26	9	9
Siddington Third	15	30	9	9
Siddington Fourth or Lowest	15	35	9	9
South Cerney Upper	16	60	9	4b
South Cerney Middle	16	65	9	4b
South Cerney Lowest	16	73	9	4b
Boxwell Spring or Shallow or Little Lock	17	45	3	6
Wilmoreway or Wildmoorway Upper or Humpback	18	1	7	6+
Wilmoreway Lower or just Wilmoreway	18	36	11	0+
Cerney Wick	19	21	6	0
Latton	20	61	9	4
Eisey	22	40	7	0
Dudgrove Double Lock	28	0		
upper chamber			9	0
lower chamber			2	6
Inglesham	28	60	6	2

* Rise as recorded in the latter part of the nineteenth century was 8 feet 5 in
\+ Fall as recorded in the latter part of the nineteenth century was 9 feet 3 in
a Recorded in 1894, almost certainly inaccurately, as 8 feet
b Recorded in 1894, almost certainly inaccurately, as 9 feet 3 inches
The distances are as recorded in the company's Fact Book

APPENDIX 2

GROSS RECEIPTS

to the nearest pound from April, 1797 to April 1881

	£		£		£
1797	4,453	1826	6,487	1855	2,874 f
1798	6,528	1827	6,441	1856	3,708
1799	5,678	1828	6,497	1857	3,445
1800	8,244	1829	6,341	1858	3,426
1801	9,831	1830	5,780	1859	3,406
1802	6,712	1831	7,519	1860	3,352
1803	6,498	1832	7,029	1861	3,709
1804	6,167	1833	7,398	1862	3,698
1805	5,768	1834	7,519	1863	3,459
1806	5,157	1835	7,470	1864	3,490
1807	5,169	1836	7,346	1865	2,610
1808	5,306	1837	7,308	1866	2,904
1809	6,249	1838	7,970	1867	2,808
1810	8,504	1839	8,574	1868	2,479
1811	7,141 a	1840	10,020 c	1869	2,590
1812	5,879	1841	11,330 c	1870	3,149
1813	5,329	1842	9,352 d	1871	2,344
1814	5,273	1843	7,199	1872	2,761
1815	6,506	1844	7,935	1873	2,435
1816	5,130	1845	8,335	1874	2,134
1817	4,915	1846	7,804 e	1875	1,812
1818	5,465	1847	6,618	1876	3,000 g
1819	6,413	1848	6,147	1877	3,730
1820	5,262 b	1849	6,553	1878	4,125
1821	6,226	1850	6,173	1879	4,150
1822	5,234	1851	5,847	1880	3,770
1823	5,272	1852	5,680	1881	3,521
1824	6,301	1853	4,291		
1825	6,495	1854	3,939		

a Opening of the Kennet & Avon Canal, towards the end of 1810, cut Bristol trade.

b Opening of the North Wilts Canal, 2 April 1819, improved the eastern outlet.
c Building of the railway between Swindon and Cirencester, brought a temporary increase of traffic.
d GWR opening to Cirencester, 31 May 1841, caused some loss of traffic.
e GWR opening to Gloucester, 12 May 1845, caused further loss.
f Conversion of the Forest of Dean Railway to broad gauge in July 1854, caused great loss.
g The Thames & Severn again became carriers in June 1875.

APPENDIX 3

THE THAMES & SEVERN CANAL COLLECTION

In order to reduce the number of notes, reference has not generally been given to the information drawn from the canal company's original books and papers in the Gloucestershire County Record Office. The collection is so vast that even a full list is unwieldy. The list appended is therefore only a selection of the more important and interesting items.

Proprietors' Registers, 24 June 1783 to 20 November 1866
Committee Registers, 23 July 1783 to 23 October 1874
Minutes of the Thames & Severn Canal Committee of the Gloucestershire County Council, 13 January 1900 to 1 July 1935
Minute Book of the Commissioners appointed under the Act of 1783
Letter Book, 1783-97
Disbursements, 1783-98
Workmens Ledger of Debits no Credits, 1783-96
Cutting Books of William Mytton and others employed in cutting the canal
Minute book of the Committee of Trade, 1794-1804
Account Book, London wharf accounts, etc., late eighteenth and early nineteenth centuries
Journal of Work done at The Bourn Yard by Samuel Bird from mid-summer 1799
Goods passed on the Thames & Severn Canal from the Rivers Thames and Severn, 1793-1804
Ledger Book No. 1
Memorandum books, late eighteenth and early nineteenth centuries, including those used by Samuel Smith and John Stevenson Salt

Transfer Register, 1797-1886

Fact Book, containing inter alia: tables of distances, locks and bridges; expense of water taken from the River Churn, 1796-1884; tonnage of coal imported, 1785-1826, of salt, 1800-26, of coal and other traffic, 1827-86; progress of the Contingency Fund, 1810-66

Book *Relating to the Steam Engine at Thames Head*, containing facts, diagrams and reports concerning the Boulton & Watt and Cornish pumping engines, the well and galleries

Reports:

Robert Whitworth to the promoters, 22 December 1782

Josiah Clowes to the committee, June 1789

John Smeaton to Lord Eliot, 14 August 1789 (copy)

Joseph Sills on the condition of the River Thames, 12 June 1794

Report Book, principally reports by Joseph Sills and James Black on matters relating to the company's trade, 1795-1804

James Black to the committee, 22 August 1797

David McIntosh to the chairman and directors, 28 June 1820

John Denyer to the committee, 18 October 1822

Final report by A. Brome Wilson, 25 March 1905

Maps and Plans:

A Profile of the Deep Cutting and Tunneling from Cotes Field, through Hayle-Wood and Sapperton Hill to Daneway Valley with the Depths to the several marks, as now Set-out. Taken in August 1783. By Robert Whitworth

Rough Plan by Jno. Doyley, Land Surveyor, bound, no date but prior to May 1789

Books of Plans, late eighteenth century

Plan of the Feeder from Jenner's Mill to the Bason, 1786

Plan by Rd. Hall & Son, Surveyors, 1810

Longitudinal Section of Sapperton Tunnel, *c.* 1857

Miscellaneous Papers including: land valuation and negotiations; agreements and tenders, 1783-93; construction of Sapperton tunnel and subsequent repairs; water supply from the River Churn at Jenour's mill and from Thames Head, 1785-1832; Boulton & Watt indenture, 1791; bills for setting up the pumping station, 1790-98; the company's financial difficulties and reconstruction, 1795-1809; legal cases; appointment of employees; bye-laws; lists of property; lists of shareholders at various dates, 1808-80; letters from John Denyer, manager, to J. S. Salt, treasurer, for nineteen of the years between 1808 and 1842; weekly return of conditions on the summit sent by the clerk-of-the-works at Siddington to J. S. Salt, treasurer, for nineteen of the years between 1809 and 1842; letters and memoranda covering the period 1876-95; abandonment of the canal.

INDEX

The more important references, including illustrations, plans, sections and diagrams, are indicated in italic type